SHAPIRO

DEVELOPMENTAL
BIOLOGY
OF THE BACTERIA

DEVELOPMENTAL BIOLOGY OF THE BACTERIA

Martin Dworkin
University of Minnesota

The Benjamin/Cummings Publishing Company, Inc.
Reading, Massachusetts • Menlo Park, California • Don Mills, Ontario •
Wokingham, U.K. • Amsterdam • Sydney • Singapore • Tokyo •
Mexico City • Bogota • Santiago • San Juan

This book is dedicated to my teachers . . .

Jackson W. Foster
Roger Y. Stanier
C. B. van Niel

Sponsoring Editor: Paul Elias
Production Editor: Greg Hubit

Copyright © 1985 by The Benjamin/Cummings Publishing Company, Inc.

All rights reserved. No part of this publication may be reproduced, stored in a retrieval system, or transmitted, in any form or by any means, electronic, mechanical, photocopying, recording, or otherwise, without the prior written permission of the publisher. Printed in the United States of America.

Library of Congress Cataloging in Publication Data
  Dworkin, Martin.
  Developmental biology of the bacteria.

  Bibliography: p.
  Includes index.
  1. Bacteria—physiology. 2. Bacterial growth. 3. Microbial differentiation.
  I. Title.
  QR84.D96   1986   589.9'03   85-15840

ISBN 0-8053-2460-7

ABCDEFGHIJ-HA-898765

The Benjamin/Cummings Publishing Company, Inc.
2727 Sand Hill Road
Menlo Park, California 94025

One can understand the essence of things only when one knows their origin and development.
—Heraclitus of Ephesus

Contents

Preface

Chapter 1 Introduction   1

    A. General Background, 1
    B. Definitions, 2
    C. The Various Categories of Procaryotic Development, 3

PART I THE ORGANISMS

Chapter 2 Regulation and Morphogenesis in Bacteriophage Development   8

    A. Introduction, 8
    B. Regulation of Phage Development, 9
    C. Phage Mu and the Strange Business of Its Inversions, Translocations, and Deletions of Segments of Host DNA, 15
    D. Phage Morphogenesis, 18

Chapter 3   The Endospore   22

       A. Introduction and Some History, 22
       B. Growth and Nutrition, 23
       C. Spore Structure, 23
       D. The Life Cycle of *Bacillus*, 30
       E. Genetics of *Bacillus*, 34
       F. Bacteriophage for *Bacillus*, 41
       G. Induction of Sporulation, 43
       H. Regulation of Sporulation, 45
       I. Spore Resistance, 48

Chapter 3   *Caulobacter*   50

       A. Introduction, 50
       B. Life Cycle, 51
       C. Ecology and Distribution, 52
       D. Taxonomy and Natural Relationships, 53
       E. Cell Structure, 53
       F. Physiological Correlates of Development, 59
       G. *Caulobacter* Phage, 61
       H. *Caulobacter* Genetics, 62
       I. Flagella Synthesis, 65
       J. Concluding Remarks and Salient Questions, 66

Chapter 5   Heterocyst Development in Cyanobacteria   68

       A. Introduction, 68
       B. The Vegetative Cell, 69
       C. The Heterocyst, 72
       D. Conclusions and Salient Questions, 83

Chapter 6   *Streptomyces*   85

       A. General Description, 86
       B. Properties of the Spore, 87

C. Spore Germination, 91
D. Antibiotics, 93
E. Genetics, 96
F. Regulation of Developmental Events, 100
G. *Streptomyces* DNA, 102
H. Extracellular Factors, 103
I. Concluding Remarks and Salient Questions, 104

Chapter 7   The Myxobacteria   **105**

A. The Life Cycle, 110
B. Ecology, 112
C. Taxonomy, 117
D. Cultivation, 118
E. Cell Structure, 119
F. Myxobacterial Polysaccharide, 123
G. Motility and Chemotaxis, 127
H. Bacteriophage, 130
I. Genetics, 131
J. Development, 136
K. *Stigmatella aurantiaca,* 146
L. Conclusions and Salient Questions, 148

PART II   THE PROBLEMS AND THE ISSUES

Chapter 8   The Relationships between Environment and Development   **152**

A. Introduction, 152
B. Formation of Resistant, Metabolically Quiescent Resting Cells, 153
C. Formation of Dispersal Mechanisms, 155
D. Formation of Reproductive Cells, 157
E. Formation of Metabolically Specialized Cells, 157
F. Formation of Optimally Efficient Feeding Cells, 158

G. Formation of Fruiting Bodies, 159
H. Conclusions, 161

Chapter 9   The Molecular Basis of Morphogenesis   162

A. Introduction, 162
B. Peptidoglycan and Cell Shape, 163
C. Flagellar Self-Assembly, 167
D. Ribosome Assembly, 168
E. *Bacillus* Endospore, 168
F. Cyanobacterial Heterocysts, 169
G. Myxobacterial Spores and Fruiting Bodies, 170
H. Lessons to Be Learned, 172

Chapter 10   Regulation of Development   175

A. Introduction, 175
B. The Timing of Developmental Events, 176
C. Spatial Aspects of Development, 177
D. Transcriptional Regulation of Development, 178
E. Translational Control of Development, 181
F. Transposable Elements and Developmental Regulation, 184
G. Epilogue, 185

Chapter 11   Exchange of Signals   186

A. Introduction, 186
B. Cyclic AMP, Chemotaxis, and Developmental Aggregation in *D. discoideum*, 188
C. Mating Signals in Yeast, 192
D. Steroid Hormones and the Water Molds, 196
E. Cell Signaling among the Prokaryotes, 197
F. Conclusions, 203

Chapter 12   Genetic Approaches to Studying Development   **204**

    A. Introduction, 204
    B. The Various Approaches, 205
    C. Transposons and Recombinant DNA Technology, 212
    D. Conclusions, 214

Chapter 13   Epilogue   **216**

References, 221

Index, 247

Preface

The study of development is one of the profound pursuits of biology and a frontier area in which new paradigms of biology are likely to be uncovered. However, until recently, bacterial development, as a field, has languished. Evidence for this is its virtual absence as a formal division at meetings for developmental biology, the lack of a journal devoted to the subject, the relative scarcity of courses in bacterial development, and the lack of appropriate textbooks for such courses. This set of circumstances is most peculiar and is in contrast to the general recognition that (a) a number of bacterial systems go through more or less complex developmental processes and (b) the experimental convenience of bacteria is always an advantage. The past two decades of molecular biology have set the stage, and various bacterial systems have been thoroughly domesticated.[1] Thus, the vigorous examination of bacterial development is now timely and appropriate.

With the above in mind, this book is designed to draw the material together in an organized fashion. As such, I intend for it to serve primarily as a text but also as a guide for the professional who is seeking an introduction to the field. As a text, it is intended for graduate students and upper division undergraduates. It presumes that the student has had prerequisite coursework in general microbiology, microbial physiology, and biochemistry.

It was necessary to make an initial strategic choice between two approaches to presenting the material. On the one hand, the subject could be presented in a strictly organismic context, with a series of chapters each covering a specific developmental system. On the other hand, the alternative was to present bacterial development as a series of conceptual areas or questions with appropriate organisms referred to when it was useful to do so. The first, or the smorgasbord approach, is the one usually employed at symposia on bacterial development, and while it preserves the organismic integrity of the field, it suffers from a conceptual disorganization and incompleteness. The conceptual approach, on the other hand, tends to subordinate the organismic richness of the field.

The resolution I have chosen acknowledges the validity of both approaches and attempts to combine them. Thus, the first part of the text covers the systems; each organism or group of organisms is described from a biological point of view. The developmental questions are put on display but not examined until the second part of the text when the questions are asked against the backdrop of organismic biology. Presumably, a course in bacterial development can be approached in the same way.

I have selected six systems for examination. These are bacteriophage, *Bacillus* endospore formation and germination, *Caulobacter* stalk and swarmer cell development, heterocyst formation in cyanobacteria, actinomycetes, and myxobacteria. This list is by no means exhaustive. There are numerous other bacteria that undergo development, and some of them have been extensively examined. For example, *Bdellovibrio*, cyst formation in *Azotobacter*, budding in *Rhodomicrobium*, bacteroid formation by *Rhizobium*, and the rod-sphere transformation in *Arthrobacter*. However, each of the six systems selected exemplifies a particular aspect of bacterial development and most appropriately illustrates the approaches being used to examine that aspect.

Finally, I have tried to avoid the temptation to write for my colleagues rather than for my students. In a sense, a textbook is a further distillation of the review articles on which it is based. It is thus inevitable and in the nature of the compromises inherent in teaching that some material will be left behind. Nevertheless, I hope that this effort will prove useful, both to colleagues and to students.

Acknowledgments

The idea for this book came from Mort Starr, who convinced me that a need existed. Without his encouragement and prodding, it would have been difficult to overcome the inertia created by the fun of doing science and teaching.

I have often been mildly amused by authors who have acknowledged the help of their spouses and families. Now I understand. There were many evenings at the family dinner table when a question or comment directed toward me was met with a blank stare, while I tried to figure out how to deal with this or that textual problem.

A number of my colleagues read portions or most of the manuscript. Their comments encouraged me to continue, and their criticisms were always helpful. These persons included Peter Wolk, Richard Losick, Jerry

Ensign, David White, John Bonner, and Hatch Echols. I owe especial thanks to my colleague Jim Zissler, who helped me to organize the ideas presented in Chapter 12 (Genetic Approaches to Studying Development.)

The illustrations were done by Randy Gooch and the typing by Patricia Graney. They did their jobs efficiently and dealt with my occasional moments of disorganization with patience and good humor.

It goes without saying that any errors in the book were either made by me or not detected by me; in either case, they are my own.

Chapter 1

Introduction

A. General Background

There are four operative factors that determine whether or not a bacterium will withstand the inexorable selective force of environmental pressure: (1) the efficiency of the organism's processes of food gathering, metabolism, and growth; (2) its genetic versatility; (3) its resistance to such environmental fluctuations as temperature, desiccation, light/dark, toxic materials, and so on, and (4) its ability to become dispersed. In the case of an organism such as *Escherichia coli*, its ability to find food chemotactically, take it up and metabolize it efficiently, grow rapidly, and adjust its metabolic processes to a frequently changing environment represents an extraordinarily effective adaptation to its environmental niche.

Accordingly, the fine details of these processes have occupied the attention of microbiologists and biochemists for the past hundred years. This preoccupation is expressed by François Jacob with typical Gallic panache: "One bacterium, one amoeba . . . what destiny could they have other than to form two bacteria, two amoebae . . .?"[1] Such an attitude emerges largely as the resolution of the controversy that raged a hundred years ago between the proponents of pleomorphism and monomorphism. The pleomorphists, usually working with mixed cultures, insisted that the form and function of microbes were extremely variable. The monomorphists, on the other hand, recognized the complications created by succession of types in mixed cultures. They formulated a conceptual as well as a technical dogma that insisted that normal, healthy bacteria had one invariable form that ideally could be best studied in exponentially growing, pure cultures.[2] The confusion generated by pleomorphists working with mixed cultures unfortunately obscured the idea that they were, in fact, partially correct. Bacteria do indeed vary in form and function; and it is probably not too far off the mark to say that their normal, natural appearance and behavior is closer to what have previously been thought of as exotic or unusual forms than to the stereotypic rod or coccus growing exponentially in a rich laboratory medium. However, powerful insights have emerged as a result of this stereotypic view; the dramatic successes of microbial biochemistry, physiology, and molecular biology testify to this as do the clarification of the etiology of most infectious diseases. Nevertheless, this view has tended to divert attention from a second evolutionary strategy—that is, development.

From a teleonomic point of view, the strategy of development allows a microorganism the advantage of temporary or permanent specialization with which to deal with a continually changing environment. This specialization may be temporal, as in the case of the formation and germination of resistant resting cells, it may be spatial as in the case of of heterogeneous filaments of cells, or it may be both as in the case of *Caulobacter*. In higher organisms, differentiation often includes the formation of specialized reproductive cells. In bacteria, however, development seems limited to providing the organisms with (a) cells that are more resistant than their growing vegetative forms (e.g., *Bacillus* endospores, *Streptomyces* arthrospores, and *Myxococcus* myxospores), and (b) processes that make food gathering or metabolism more efficient (e.g., myxobacterial swarms, *Caulobacter* stalks and swarmer cells, and cyanobacterial heterocysts). In any case, the strategy of development, as a counterpoint to that of maximizing the growth rate, may be a far more common strategy in nature than has previously been recognized.

B. Definitions

A definition of development can be useful from the points of view both that it will tend to delimit and organize the subject matter and that it might lead to sharpening one's thoughts about the process. Definitions of development sometimes tend to founder along the way to *reductio ad absurdum*. That is, unless certain important constraints are made, any metastable physiological or biochemical process becomes a developmental one and the phenomenon comes to include everything and therefore nothing. For example, François Jacob and Jacques Monod in the flush of success in 1963 over the discovery of gene regulation suggested the following definition of cell differentiation: "Two cells are differentiated with respect to each other if, while they harbour the same genome, the pattern of proteins which they synthesize is different."[3] And Joshua Lederberg in 1966, in an attempt to demystify development and purge it of its reliance on amorphous organizers and inducers, suggested that one could reduce development essentially to differential gene expression.[4] Let us instead try the following: "Development is a series of stable or metastable changes in the form or function of a cell, where those changes are part of the normal life cycle of the cell." The obvious question is "What is included in 'the normal life cycle of the cell' "? Is the formation, spacing, and timing of a division septum part of the normal life cycle? Is, indeed, the induction of β-galactosidase by *E. coli* part of the normal life cycle? Is the alternation between phototrophic and chemotrophic growth in *Rhodopseudomonas* part of the life cycle? The latter certainly involves profound changes in the cell's physiology albeit not in its outward form. And it is, like sporulation, a substantial change in the cell's modus vivendi in response to a shift in the environmental conditions.

All these changes are, in the strictest sense, developmental ones; yet they fail to conform to my intuitive sense that a substantial change in form as well as function defines the developmental process. Discussions of phage morphogenesis have pointed out that development is a process involving three features: (1) there is a choice — a branch point that leads to (2) a pathway or a series of events; (3) to some extent the cell is locked into the pathway — that is, there is a point of no return.[5] To this I would add (4) the pathway results in a substantial change in the form and function of the cell.

There is an alternative approach for trying to surround the concept of development with a definition. Rather than trying to define development in terms of process, one may define it in terms of function. Thus, returning to the idea presented at the beginning of this chapter, development may be defined as a strategy such that the organism, rather than consisting of one essential cell type that deals less than optimally with a wide variety of environmental circumstances, consists of multiple cell types each limited in versatility but highly adapted to a particular circumstance. In a sense, cellular specialization is substituted for versatility. One may add to this the fact of cellular differentiation — that is, the idea that the cell itself may become asymmetrical and functionally differentiated.

C. The Various Categories of Prokaryotic Development

Where attempts have been made to categorize prokaryotic development, these have usually been in terms of morphogenesis and differentiation. The former is used to refer to "changes in external morphology of the cell and in internal architecture during the cell cycle"[6] and the latter to "events initiated by a 'switch' in the cell cycle leading to the formation of a new type of cell."

T. M. Sonneborn[7] has referred to these processes as intracellular and intercellular differentiation, respectively, and has pointed out that "every occurrence of . . . *inter*cellular differentiation is accompanied by differentiation within one or both or all of the divergent cells—that is, by *intra*cellular differentiation." He goes on to point out that we therefore study morphogenesis in order to understand differentiation.

It is useful to point out another fundamental dichotomy among the various developmental processes. Development may result in a temporary state of specialization that allows the cell to cope with either changing environmental conditions or with a transient phase of an organism's life cycle; or it may be a series of changes leading to a stable, dead-end cell. For example, the future of the germ cell in multicellular eukaryotes or of the bacterial spore includes a reversal of the differentiated state and a return to the alternative growth mode. On the other hand, somatic cells of multicellular eukaryotes or heterocysts of cyanobacteria serve their particular function and are then developmentally doomed. In one case—the spore—development is reversible—that is, the process is cyclic. In the other case, the process is unidirectional, and the cell will normally not reproduce. This point is not merely an abstract one but may have specific consequences; intuitively, one would imagine that the nature of the regulatory devices at play in both these processes would differ substantially. It is not yet clear whether or not this is indeed the case.

Roger Whittenbury and C. S. Dow[6] have pointed out yet another useful dichotomy in prokaryotic development. They distinguish between cell cycle–independent differentiation as exemplified by endospore formation in *Bacillus* and cell cycle–dependent differentiation illustrated by the dimorphic life cycle of *Caulobacter*. In the latter case, development is complexly interwoven with growth and continued DNA replication. As will be discussed in more detail in Chapter 4, the stalked cell of *Caulobacter* acts as a stem cell, continually giving rise by growth and asymmetric division to daughter swarm cells. It is obvious that the regulatory strategies involved in these two kinds of developmental processes are likely to be quite different.

A final distinction can be made between unicellular and multicellular or colonial development. What has been discussed until this point has fallen in the former category and refers to the changes in form/function undergone by a single cell. However, many bacteria exist as organized populations and manifest a rudimentary but organized multicellularity. Examples are the formation of fruiting bodies by the myxobacteria, heterogeneous filaments and aggregates of cyanobacteria, and the mycelial mat of actinomycetes. These processes involve tactic, developmental movement, the exchange of chemical signals between cells, the coordinated construction of complex, multicellular structure, and most likely, cell-surface interaction between cells. This area of bacterial development is extremely rich and relatively unexplored.

Obviously, the existence of a differentiated population of specialized or dead-end cells coexisting with vegetative or reproductive cells requires that the various populations are constrained to remain within a physical boundary, (i.e., are multicellular). Thus within the prokaryotes, one would expect to find population differentiation and unidirectional development only among those organisms that are filamentous (e.g., *Anabaena, Streptomyces*) or those that formed organized, adhering masses of cells (e.g., the fruiting bodies of myxobacteria. On the other hand, in the case of those prokaryotes that undergo development but are essentially unicellular organisms—*Bacillus*, for example—one is not surprised to discover that development (i.e., spore formation, spore germination) is cyclic and metastable. (In *Caulobacter*, where the stalked cell represents a kind of primitive stem cell perpetually giving rise to flagellated swarm cells, the neat dichotomy starts getting compromised.)

It seems useful now to summarize the various distinctions that have been made thus far, by means of the following simple taxonomy of development:

1. Unicellular development
   a. Cyclic development, e.g., the formation and germination of spores and cysts
   b. Noncyclic development
      1. Dead end cells, e.g., cyanobacterial heterocysts
      2. Stem cells, e.g., stalked cells of *Caulobacter*
2. Multicellular development
   a. Aggregates generated by developmental movement, e.g., fruiting bodies of myxobacteria
   b. Aggregates generated by oriented growth, e.g., actinomycetes
   c. Heterogeneous filiments, e.g., vegetative cells, heterocysts, and akinetes in cyanobacteria

At this point I wish to define in a somewhat more formal sense some of the phrases that are frequently used to describe the various aspects of development.

1. *Cellular morphogenesis* refers to those processes involving the actual changes in form of the cell. The formation and germination of the various spores, cysts, and akinetes exemplify this process. The process may either be cyclic as in the case of endospores or unidirectional as in the case of heterocysts.
2. *Cellular differentiation* refers to those processes of decision whereby, as a result of a morphological change in part of a cell, those parts acquire a different function from the rest of the cell. Examples of this are acquisition of stalks by *Caulobacter*, of unipolar pili by *Myxococcus*, or the differentiation of a *Bacillus* vegetative cell into an endospore and a sporangium.

3. *Population differentiation* refers to those processes that lead to the generation of functional heterogeneity among a population of cells. The presence of akinetes, heterocysts, and vegetative cells along a filament of *Anabaena* or the coexistence of hyphal spores and aerial and substrate mycelia in *Streptomyces* are clear examples.

In other words, *differentiation* should be used to refer to those decision-making processes that will lead to heterogeneity either at the cellular or population level. *Morphogenesis*, on the other hand, is intended to refer to the subsequent, actual process of change in shape or function.

4. *Cell communication and interactions.* Most of us were taught that bacteria are unicellular creatures—that is, that the properties of a bacterial population are exactly the sum of the properties of the individual cells. It has become increasingly clear that this is not the case. In some bacteria (perhaps far more frequently than we imagine), cells are interacting and communicating with each other in a varied and systematic fashion. The two systems where this has been examined are in the heterocyst-vegetative cell interactions in cyanobacteria and, more intensely, during growth and development in myxobacteria. Thus, this phenomenon includes those situations where the behavior or development of a cell is affected by contact with or a signal from another cell.

Finally, some attention will be paid to analyzing the processes of self-assembly during bacteriophage synthesis. While this process does not fall in the realm of bacterial development, the principles it exemplifies are certain to play a role in bacterial morphogenesis. Moreover, the working out of the details of bacteriophage assembly was a tour de force and deserves to be admired and relished, if only for its own sake.

It shall be the theme of this book that there is a central core of questions that the microbial developmental biologist is interested in and that most of those questions can be asked either in the context of prokaryotic or eukaryotic microbes. It is not my goal to make a case for one or another system; rather, I shall point out the kinds of developmental questions that can be examined in prokaryotes. A second theme shall be that one or another particular prokaryotic system may be most suitable for examining a particular kind of developmental question.

I hope it will be obvious that the molecular and genetic approaches, against a backdrop of a broad biological understanding of the organism, are now available, and both approaches are *sine qua non*. Whether these will in fact come together to generate that deep and broad understanding of development that is at the core of biology is finally a matter of faith.

PART I

THE ORGANISMS

Chapter 2

Regulation and Morphogenesis in Bacteriophage Development

A. Introduction

There are numerous reasons why any serious student of developmental biology must be familiar with the details and strategies of bacteriophage development. I shall list these reasons and then return to discuss each of them in more detail.

First, phage development embodies many of the components of the definition of development presented at the outset. In the case of a lysogenic phage, there is a branch point, followed by a pathway or chain of events, culminating in a morphogenetic event. The decision at the junction of lysogeny and vegetative development is analogous in a formal sense to the decision between growth and development of a bacterial cell. Second, there are a variety of mechanisms for regulating transcription of the phage genome examplified by the phages T7, SPO1, and λ. These mechanisms may turn out to reflect similar strategies to those employed in developmental operons, if such indeed exist. Third, certain bacteriophage, notably the *Escherichia coli* phage Mu, have revealed a bizarre and idiosyncratic behavior of their DNA resulting in fusions, transpositions, inversions, or deletions in the cellular DNA, any of which may, if carried out in a regulated fashion, play a role in developmental regulation. Fourth, the self-assembling properties of phage proteins suggest a way of thinking about certain aspects of cellular and colonial morphogenesis that has otherwise defied analysis.

The examination of these processes, along with a more general recognition of the role of transposable elements or "jumping genes," has opened up thinking about mechanisms of genetic change. It may turn out that a somewhat more modulated or regulated aspect of Mu behavior, such as found in Mu *gin* (see Section C of this chapter) or of site specific recombination in λ phage, may provide a model for the kind of metastable genetic change needed to regulate developmental processes.

B. Regulation of Phage Development

1. Transcriptional Control

Before addressing the questions I have posed, refer to Figure 2-1 to refresh your memory as to the general outline of phage infection and replication. The phage attaches to the cell surface by means of an attachment mechanism and specific receptor sites on the cell surface. The phage nucleic acid is injected into the host cell and the subsequent proper and orderly development of additional phage particles depends on the proper and orderly transcription of the phage genes and some additional posttranscriptional controls. If the right gene products are made at the right time, their orderly replication will occur, and phage components will undergo properly timed spontaneous assembly into a phage particle. Thus, the problem of regulation of phage development largely becomes one of the carefully programmed regulation of transcription of host and phage genes. The various strategies used to accomplish this regulation are as follows:

a. New RNA polymerase.[1] Phage T7 belongs to a group of phages called the T-odd phages, is infective for the coliform bacteria, and contains about 26×10^6 daltons of DNA. In the case of phage T7 infection, transcription of the phage genes is carried out both by the host polymerase and by a completely new polymerase coded for by the phage genes. During the 30-minute period between infection and lysis of the host cell, three classes of phage proteins are synthesized. These classes are transcribed sequentially from one end of the phage chromosome to the other and consist of Class I proteins (4–8 min after infection), Class II proteins (6–15 min after infection), and Class III proteins (7–19 min after infection). These proteins are also referred to as *early, middle,* and *late* gene products. The early genes are transcribed by the host cell polymerase, which is able to recognize three, closely spaced promoter sites on the T7 gerome. There is a single termination site (which does not require the termination factor rho), and arrival at this site represents the end of host polymerase function. At 4 to 6 minutes

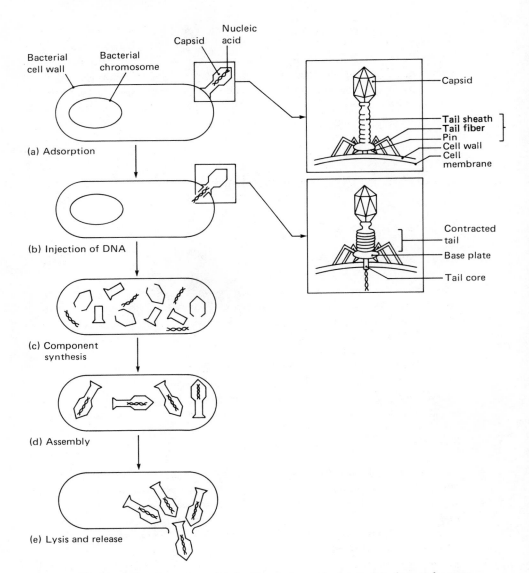

Figure 2-1 Growth cycle of a T-even bacteriophage: (a) phage adsorption; (b) injection of phage DNA into the host cell; (c) synthesis of phage DNA and capsids; (d) phage assembly; (e) host cell lysis and release of free phage. (Adapted from Tortora et al. 1982)

after infection, a T7-specific polymerase is synthesized that is then capable of transcribing the Class II and Class III phage genes. This polymerase is distinct from the host polymerase, differing in structure and specificity. These events are represented in Figure 2-2, which is a genetic map of bacteriophage T7. It is interesting and perhaps relevant to developmental problems that the sequential transcription of the phage genes results in a temporal sequence of events.

b. Antiterminator.[2] During lytic growth of the E. coli phage λ, a repressor protein that, in the lysogenic cell, is normally bound to two promoter sites on the λ genome is removed. In a cell that has not been lysogenized by λ, the repressor protein is, of course, absent. Removal of the repressor protein from the promoter sites P_r and P_1 allows the unmodified host polymerase to transcribe two genes called N and Cro. The product of the N gene, referred to as N protein, causes the polymerase to ignore the normal termination signal. Thus the polymerase reads through the normal stop sign and transcribes the subsequent genes necessary for phage production. The product of the Cro gene prevents transcription of the C_I gene that codes for repressor and thus helps ensure a lytic response in those cells in which the lytic pathway is selected; Cro also facilitates late lytic development in other ways.

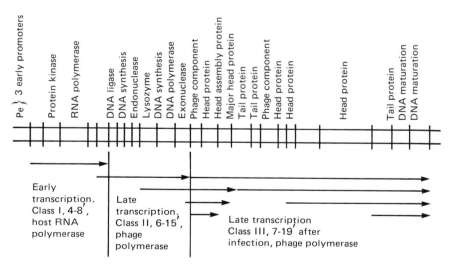

Figure 2-2 The genetic map of bacteriophage T7 indicating the temporal division of transcription into three classes. The "early" promoters are indicated by Pe; the position of the late promoters is uncertain. The extent of the various polycistronic mRNA's is indicated by the arrows. (Adapted from Mandelstam et al. 1982)

c. Positive RNA polymerase modifications.[3] Phage SPO1 is a large, virulent phage that attacks the Gram-positive bacterium, *Bacillus subtilis*. Unlike phages λ and T7, SPO1 infection induces the host cell to synthesize a series of peptides that change the transcriptional specificity of the host RNA polymerase. While there is disagreement about the exact number and nature of these peptides, there seem to be three major subunits that are coded for by the phage, associate with the host RNA polymerase, and change its transcriptional specificity. During the early stage of phage infection (0–4 min) transcription of the phage genome is carried out by the unmodified host polymerase. One result of this transcription is the synthesis of a 28,000-MW peptide that, when bound to the host polymerase, allows it to transcribe the middle genes (4–8 min). Two additional peptides are synthesized (MW = 13,500 and 24,000), which then permit transcription of the late genes. The evidence is not yet compelling but does suggest that the altered specificity of the polymerase is attributable to the peptide-induced modifications of the ability of the polymerase to recognize specific promoter sites. Ordinarily, transcriptional specificity in *Bacillus* is conferred upon the polymerase by the peptide σ. In SPO1-infected cells, the σ factor is absent and is replaced by the phage-induced peptides. This phenomenon is a clear and direct example of positive regulation by phage-induced proteins that direct transcriptional specificity by binding to RNA polymerase.

d. Negative RNA polymerase modifications.[4] During the early stages of T4 and T7 phage infection, some of the host functions of the RNA polymerase are inhibited. This inhibition occurs specifically in the case of the T4 phage of *E. coli* where the α, β, β' subunits are modified by the addition of ADP-ribose to specific arginine residues in the host RNA polymerase. The donor molecule for the ADP-ribosylation is oxidized NAD; phage mutants lacking the ability to carry out some of these polymerase modifications have been isolated and seem able to undergo normal development and expression of phage genes. No mutant of T4 deficient in all of these modifications has yet been isolated. Thus, while it appears that the modification may play some role in inhibiting host functions, the issue is not completely resolved.

A similar function is expressed in cells of *E. coli* infected with phage T7. In this system, the transcription of host genes is shut off at 3 minutes after infection and that of early phage genes at 8 to 12 minutes after infection. This is correlated with the appearance of a kinase, the gene product of an early gene (0.7), which has the ability to phosphorylate the β' subunit of the polymerase. This function seems to be necessary for effective phage propagation under conditions of poor nutrition or high temperature.

Finally, T7-infected cells produce a 14,000-dalton polypeptide that specifically inactivates host holoenzyme but not core RNA polymerase. The inhibitor binds to the polymerase and may also act to prevent transcription of genes no longer relevant to phage production.

Thus, a number of logical possibilities for altering transcriptional specificity seem to have been discovered and employed by the various phages. These possibilities include synthesis of a new polymerase, modification of the host polymerase to allow developmental transcription or to inhibit nondevelopmental transcription, and modification of the transcriptional signals on the DNA. It is not surprising (as shall be discussed in some later chapters) that there are certain analogies with the transcriptional regulation in some developing bacteria.

2. The Choice between Lysogeny and Productive Infection[5]

A temperate phage is one that can reproduce by two alternative methods: it may either undergo a productive infection, lyse its host cell, and then infect another cell nearby, or it may be transmitted from mother to daughter cell without an intervening extracellular stage. In the latter case, the most common mechanism is for the phage genome to be physically integrated into the genome of the host. Replication of the host genome thus includes the phage and guarantees its transmission from one generation to the next. This mode of establishing a stable lysogenic complex is exemplified by the relationship between E. coli and bacteriophage λ. An alternative mode is manifested by bacteriophage P1, which exists in its lysogenic state, not as an integrated part of the host chromosome but as an independent, extrachromosomal plasmid. Regardless of the mechanism by which the temperate phage establishes the lysogenic state, it must accomplish two things. It must guarantee that with each replication of the host genome there is a faithful replication of the phage genome, each daughter cell then receiving a copy. It must also turn off the expression of the phage genes whose transcription and translation would lead to phage replication and disruption of the host cell. This arrangement, however, is metastable; there is a certain low frequency with which the lysogenic association is terminated and productive infection ensues.

The situation is thus precisely analogous to those one would traditionally describe as developmental; a metastable commitment to one of two alternative sequences of events leading to morphologically distinguishable consequences.

The association between E. coli and the temperate phage λ has been intensively studied and is understood in greater detail than any other biological interaction. It is thus useful to examine this interaction as a prototypic developmental process. As indicated earlier, there are two aspects of the lysogenic process, both controlled by λ genes—the faithful transmission of the phage genome from one generation to the next and the suppression of those functions that would lead to phage reproduction. Although it turns out that in λ the two processes are connected, we shall focus our attention on the latter.

Lambda phage is smaller than the T-even phages but shares with them the property of a long tail and a polyhedral head. Lambda DNA exists in the phage as a linear molecule; however, upon infection of a sensitive *E. coli*, the DNA circularizes. At that point, whether the infection becomes lysogenic or productive depends on the complex interplay between the regulation and products of the two phage genes, cI and Cro. The product of cI is referred to as λ repressor; the product of the Cro gene is referred to as Cro. Lambda repressor prevents transcription of those early genes necessary for phage growth; Cro is necessary for phage growth. Overall regulation of phage growth or lysogenization depends on the interaction of these two proteins with three operator sites. These interactions then determine whether the proteins necessary for phage growth are synthesized or whether λ repressor continues to be synthesized.

Lambda repressor not only prevents the lytic cycle by turning off transcription of early phage genes including Cro it also is required for transcription of cI, leading to its own synthesis. (This phenomenon is referred to as an *autogenous circuit.*) This stimulation of cI transcription is prevented by Cro, which at high enough concentration will turn off its own synthesis. It is also prevented by the absence of its own gene product. Thus, there is a paradox in the situation necessary for establishing lysogeny. How does cI gene product get there in the first place? The answer is that cI transcription is a secondary event; it is a consequence of the prior transcription of two other genes, cII and cIII. Their gene products circumvent the need for the cI gene product and allow the transcription of cI to be turned on by this alternative route. When both cI repressor protein and Cro gene product are finally present, each of these regulatory proteins antagonizes the synthesis of the other as well as regulating its own synthesis. These feedback loops could result in oscillatory changes in the two molecules; if these oscillations are out of phase, then the combined effects of the two molecules would be magnified. Some of these regulatory interactions are illustrated in Figure 2-3. It is an especially nice model for a developmental regulatory circuit where an initial regulatory decision may or may not be confirmed by subsequent events.

Another key phage gene in the process is gene N, whose protein product is required for the transcription of those genes responsible for somewhat later functions. N protein acts as an antiterminator, preventing the ϱ-mediated termination at sites just downstream from gene N. There is some evidence to suggest that the mechanism whereby this positive control protein interferes with termination may involve its ability to bind to the *E. coli*

RNA polymerase and thus to modify the ability of the polymerase to recognize the termination site. The products of yet another regulatory gene, Q, are necessary to permit the synthesis of the structural proteins of the virion as well as the proteins needed for cellular lysis. The gene product of Q, like that of N, functions as an antiterminator.

It seems quite clear that the complex regulatory network that controls the λ–E. coli interaction is a superbly instructive paradigm for approaching the intricate tapestry of regulatory interactions involved in development.

C. Phage Mu and the Strange Business of Its Inversions, Translocations, and Deletions of Segments of Host DNA[6]

Phage Mu (for "mutator") is a temperate phage that infects E. coli. The extraordinary feature of Mu is that the lysogenic state does not involve the insertion of the phage genome (MW ≈ about 25×10^6 daltons) at a particular, fixed site; rather Mu becomes inserted in the host promiscuously, generating a mutation wherever it has intercalated into, and therefore interrupted, a host gene. In other words, Mu acts not only as a phage but also as a transposon. The insertion of Mu is almost, but not quite, random.

There is a set of further consequences of lysogenization that distinguishes Mu from other temperate phages and emphasizes its commonality with transposons; the host chromosome is induced to undergo fusion with other unrelated circular DNA's in the cell, such as λ gal or F'lac, as well as to generate other genetic rearrangements such as inversions, transpositions, and excisions. While the exact mechanism whereby these changes in the host chromosome are induced is not known, it seems clear that they are in fact controlled by the Mu genes controlling replication and intregration of Mu DNA.

Another intriguing feature of Mu which has striking relevance to the problem of regulation of developmental events is the presence in the Mu genome of a 3,000 base-pair piece of DNA called the G segment. However, before describing the properties of the G segment, it is useful to set the stage by a brief digression concerning recent insights into the regulation of flagellar phase variation in Salmonella.

Bacterial flagella are composed of a single protein, flagellin, whose molecular weight and precise properties vary from species to species. In Salmonella, however, two types of flagella are alternatively produced, resulting in two antigenically distinguishable flagellar types referred to as H1 and H2. Cells undergo transition from one type to another with a high frequency, varying from 10^{-3} to 10^{-5} per cell per generation. In a masterful piece of work from Melvin Simon's laboratory in California,[7] it has been

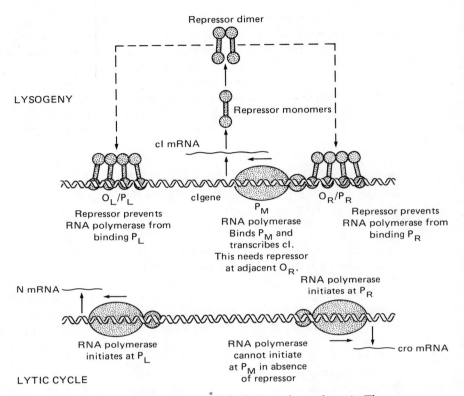

Figure 2-3 Control of lysogeny and the lytic cycle in phage λ. The upper, autogenous circuit controls lysogeny; interruption of this circuit leads to the lytic cycle (lower circuit). (Adapted from Lewin 1983)

shown that this variation is a reflection of the orientation of a 900 base-pair piece of DNA; in one orientation the structural gene for one flagellar type is expressed, as is a gene whose product prevents transcription of the gene coding for the alternative flagellin. In the reverse orientation, neither the first flagellar gene nor the gene coding for the repressor is expressed; the second flagellar type is thus expressed. The inversion of this 900 base-pair segment is controlled by a 500 base-pair gene within the invertible segment (called *hin* for h inversion). Since the *hin* gene can act in trans, it is presumably producing a diffusible substance whose function it is to mediate inversion of the entire 900 base-pair sequence. This model is diagramatically illustrated in Figure 2-4. The nature of this material has not been determined, nor is it clear what regulates the frequency of the inversion. However, this model has provided a prototype for the kind of metastable regulatory device that will most certainly also play a role in development.

Returning to phage Mu, if one examines the phage DNA by means of electron microscopy, it turns out that there is a section of the genome that

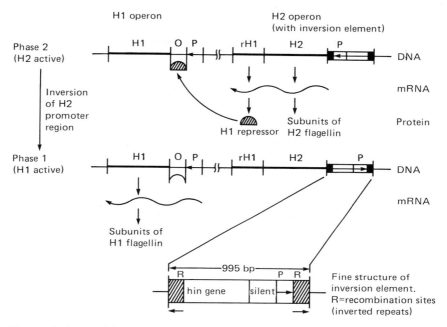

Figure 2-4 Model for the mechanism of flagellar phase variation in *Salmonella*. The diagram illustrates the two states of flagellar phase variation. In phase 2 (upper portion) both H2 and rH1 are transcribed resulting in the production of H2 flagellin and repressor protein. The latter binds to the operator *(O)* of the H1 operon and prevents synthesis of H1 flagellin. In phase 1, the *hin* gene within the promoter *(p)* is inverted and only the H1 gene is transcribed. (Adapted from Nover at al. 1982)

undergoes occasional inversion. This phenomenon was revealed by an electron microscopic technique called "heteroduplex mapping." It is based on the fact that, if there is a segment of the DNA that occasionally undergoes a flip-flop and changes its orientation, upon denaturation and reannealing with other strands that have not undergone the flip-flop, there will be areas of the DNA that no longer base-pair and are revealed as "bubbles." (Figure 2-5).

It appears that the orientation of this region, called the G segment and consisting of 3,000 base pairs, controls the ability of the phage to adsorb to a sensitive host. In one orientation, *flip*, the phage can adsorb to one particular host; in the other orientation, *flop*, the phage synthesizes a different set of tail fibers and adsorbs to a different host. The inversion is controlled by a gene called *gin* (for G inversion) which, unlike the analogous *hin* gene, is located outside the invertible sequence itself. One final remarkable fact: the unrelated coli phage P1 contains a segment of its DNA whose base sequence is identical to that of the G segment of Mu. Unlike Mu, P1 does not

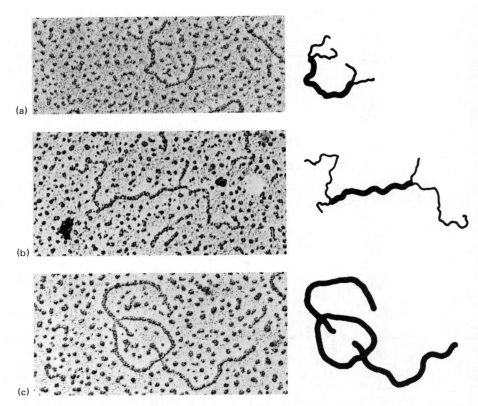

Figure 2-5 Diagrams and electron micrographs of heteroduplexes of plasmid DNA. *A* and *B* represent the annealing of two different sets of homologous strands of DNA. *C* illustrates the pairing of the homologous portions of two strands and the "bubble" where the non homologous segments have failed to pair. Both electron micrographs and corresponding diagrams are presented. (Courtesy of Dr. M. Simon)

integrate at all but exists in its lysogenic state as an extrachromosomal, independent plasmid.[8]

It is quite clear that there is a remarkable flexibility of the location and orientation of certain genes which adds a dimension to the ability of organisms to generate a variety of metastable regulatory states.

D. Phage Morphogenesis[9]

One of the most challenging and difficult problems of development concerns the process of morphogenesis. More explicitly stated: (1) What is the molecular basis of shape? What molecules and what properties of these molecules are critical for specifying the shape of a cell or an organelle? (2) How is the process ordered and organized?

Up to now our discussion of the regulation of phage development has focused on temporal strategies modulated by feedback controls. Make the right product in the right amount at the right moment, and it will be properly integrated into the complex mosaic of dynamic events.

But this tells us little about morphogenesis per se. And here the process of self-assembly first described with bacterial flagella and later in ribosomes and phage may prove instructive. The word *self-assembly* simply refers to the notion that the information for the assembly of a higher-order structure than a molecule (e.g., a multimolecular enzyme complex, a micelle, an organelle, or a virus) is contained in the primary structure of the component molecules themselves; that no additional organizational principles, enzymes, or templates are necessary, but rather that the molecules will spontaneously and independently assemble in a thermodynamically stable configuration. The paradigm for this idea has been the description of the morphogenesis of T4 phage. As a result of this elegant and remarkable work, it is now possible to describe the morphogenesis of an organism in complete molecular detail. As such, it may offer a conceptual handle for thinking about some aspects of cellular morphogenesis, while keeping in mind that cellular growth and viral assembly are fundamentally different processes.

The manufacture of the T4 phage proceeds along four essentially independent assembly lines. These lead to the formation of the phage head, the base plate, the tail fibers, and the tail itself. The proteins for each of these subunits are present in the infected cell essentially concomitantly; however, the assembly of each subunit involves a linear dependent sequence of events. Thus, each protein is used when the subunit assembly has arrived at the stage where that protein will "fit" the assembly. Certain proteins are used to form a scaffold or framework for the head. When the scaffold has served its purpose, the proteins in its assembly are hydrolyzed by a phage-induced protease.

These insights into the process are the result of two experimental strategies. In the first, morphological mutants of the phage were generated and the stage at which development was blocked could be correlated with the absence of a particular protein known to be present in the intact phage. In the second, an in vitro assembly system was developed so that complete phage could be assembled in the absence of intact cells. For example, if an extract of cells infected with a mutant phage that could generate heads but no tails was mixed with an extract containing a mutant that could generate tails but no heads, intact phage were assembled. This complementation assay could even be used to examine a cell extract for the presence or absence of a single specific protein. Thus, it has been possible to determine that of the 140 known genes in T4 phage, 55 are required for phage assembly. The assembly process is illustrated in Figure 2-6.

Nothing yet has been said about the process whereby phage DNA is incorporated into the phage. Intuitively, one might guess that the phage head is simply built around the DNA; this is not, in fact, what happens. Empty phage heads are first synthesized. At the same time, the DNA is synthesized as long concatamers, with many copies of the genome linearly

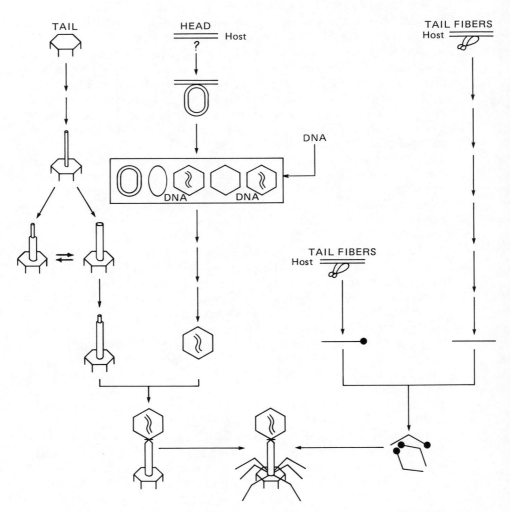

Figure 2-6 Diagram showing the pathways for the assembly of phage T4. The three pathways represent the concurrent assembly of tail, head, and tail fibers and their final assembly into the complete virion. (Adapted from Mandelstam et al. 1982).

arrayed in tandem. These are then cut by an endonuclease so that each packaging unit is slightly larger than the entire genome. Thus, for phage T4 each unit will begin at a different gene, will end up at a different gene, and will be slightly longer (about 2%) than the complete genome. Other phages, such as λ, also package from concatemers but produce uniquely terminated ends. This process is linked in some way to the uptake of the DNA so that the DNA is sucked into the phage and then ordered into the proper configuration. The mechanisms by which this occurs and the nature of the DNA-protein interactions that are involved are, in many details, unclear. The other conceptual problem that remains to be clarified is how the length of the tail or the size of the head is determined and limited.

Chapter 3

The Endospore

A. Introduction and Some History

The bacterial endospore was first described, essentially simultaneously in 1876, by Ferdinand Cohn and Robert Koch.[1] Cohn's interest stemmed from the continuing controversy over spontaneous generation, and his description of heat-resistant spores solved the problem of the ability of infusions that had been heated to 100 C to give rise to turbid suspensions of bacteria. Koch's interest was in the etiology of anthrax, and his observation of spores of *Bacillus anthracis* emerged as a result of his careful description of the life cycle of the organism.

The ensuing hundred years have seen a sustained, continuing interest in the properties of endospores as well as in the developmental events leading to their formation and germination. This interest reflects not only the practical consequences of their heat resistance (e.g., in canned foods) but also the intriguing problem of the mechanism of resistance. It was inevitable that these interests should lead to a concern as to the regulation of sporulation and germination, and indeed this area has been preeminent in prokaryotic development. Despite the awkward fact that a full understanding of the process continues to elude us, there is more known at a molecular level about sporulation than about any other developmental event.

The formation of resistant resting forms of cells is widespread among bacteria, occurring in both Gram-positive and Gram-negative cells. Among the Gram-positive bacteria, one type of spore is called an *endospore*, so-called because it is formed within a cell rather than by conversion of the entire cell to a spore. In fact, it has been suggested that the early events in endospore formation represent a kind of asymmetric cell division with one daughter cell becoming a spore and persisting, and the other representing a dead-end that eventually dies and lyses. For at least a period of time each cell contains a functional genome, and the interaction between these two genomes is an interesting and unresolved feature of the process.

There are numerous Gram-positive bacteria that form endospores; of these the genus *Bacillus* has received the most attention and shall thus be the object of our attention (Figure 3-1). These are aerobic organisms, rod-shaped, from 2 to 7 μm long and 0.3 to 1.2 μm in diameter. They are usually motile, and when so, by peritrichous flagellation. Their range of G+ C% values varies from 32 to 62%.

B. Growth and Nutrition

The nutritional requirements for vegetative growth of members of the genus *Bacillus* are extremely varied. While all are aerobic chemoheterotrophs, some such as *B. subtilis* or *B. megaterium* will grow on a simple, defined medium containing glucose and a few salts. Some have extremely complex requirements while others have not yet been cultivated in artificial media.

B. subtilis grows with a doubling time of about 30 minutes in a complex medium and about 45 minutes in a minimal, glucose-salts medium. In either case, sporulation occurs after exponential growth has ceased (Figure 3-2) and may, in fact, be reversed if fresh nutrients are added to the medium. It thus seems quite clear that the process is under nutritional control, which will be discussed in greater detail at a later time.

C. Spore Structure[2]

All endospores that have been examined thus far are composed of six parts:

1. The Spore Protoplast or Core

The part of the spore analogous to the cytoplasm of the vegetative cell, the spore protoplast or core contains the soluble enzymes of the cell, the

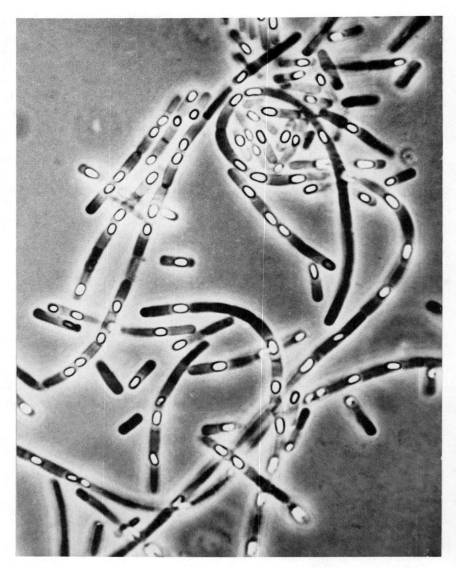

Figure 3-1 Phase-contrast photomicrograph of spores and vegetative cells of *Bacillus fastidiosus*. (Courtesy of Dr. S. C. Holt)

biosynthetic apparatus, the nucleotides, coenzyme, and metabolite pools, and the spore nucleus. Most investigators agree that the spore contains only one, completed chromosome[3]. The figure for the size of the chromosome depends on the method that is used to determine it and in vegetative cells of *B. subtilis* varies from 1.8 to 4.3 $\times$ 10^9 daltons. However, the most recent

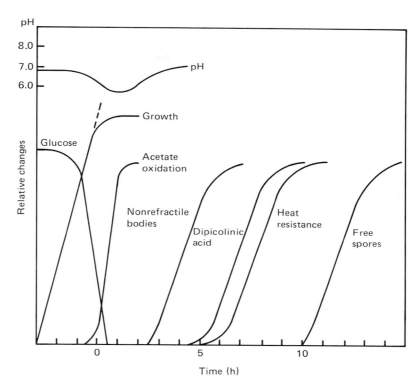

Figure 3-2 Graph showing the time of appearance of various physiological and morphological events during growth and sporulation in *Bacillus*. (Adapted from Mandelstam and McQuillen 1968)

review on the *B. subtilis* chromosome concludes that its size is 2.0 to 2.5 × 10^9 daltons, approximately the same size as the *Escherichia coli* chromosome. It is interesting that the additional information necessary for the formation and germination of the endospore does not contribute that much to the size of the chromosome.[4]

The spore is characterized by an extremely high concentration of Ca^{++} (1 to 3% of the spore dry weight) and a large amount (10% of the spore dry weight) of a unique compound called dipicolinic acid (DPA) (Figure 3-3).[5] A variety of different approaches have shown conclusively that these compounds (which probably exist in a chelate complex with each other) are located in the spore core. The role of the CaDPA in relation to the peculiar properties of the spore is unclear and this will be discussed in more detail later.

Finally, the spore core contains several low molecular weight proteins that represent 40 to 50% of the total proteins present in the core. These

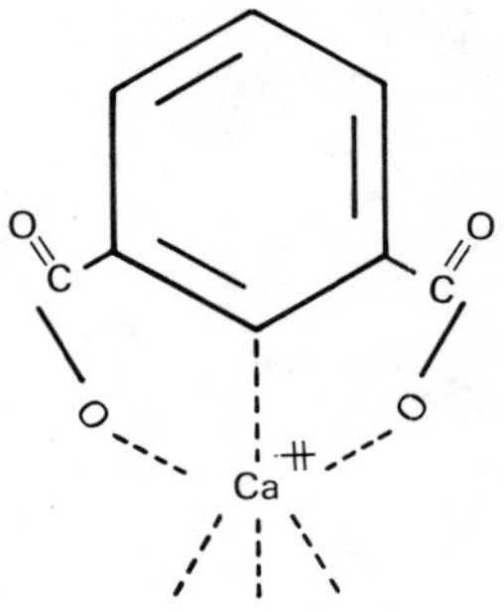

Figure 3-3 Chemical structure of calcium dipicolinate.

proteins, together with a set of proteolytic enzymes that seem to be specific for the three proteins comprise a system that, during the early stages of germination, can rapidly provide a source of amino acids that serve both as a source of energy and biosynthetic precursors.[6] Figure 3-4 is an electron micrograph of a thin section of an endospore and illustrates many of the structures that will be referred to in the following sections.

2. Inner Forespore Membrane (IFsM)

The spore protoplast is surrounded by plasma or spore membrane. Its composition seems to be similar to that of the membrane of vegetative cells; and it seems reasonable to assume that its functions are essentially the same as those of the vegetative cytoplasmic membrane—namely, as a boundary for the spore protoplast, as a regulator of cell permeability, and as a structural matrix for certain of the cell's metabolic enzymes (e.g., the electron transport system).

3. Germ Cell Wall

Immediately external to the plasma membrane is a structure called the germ cell wall. This structure seems to consist solely of vegetative cell peptidoglycan (unlike the peptidoglycan in the spore cortex which is chemically quite different), and it has been suggested that upon spore germination, it becomes once again the vegetative cell wall. Unlike the cortex peptidoglycan, the germ cell wall is synthesized from precursors provided by the spore protoplast and transported through the plasma membrane.

Chapter 3 The Endospore 27

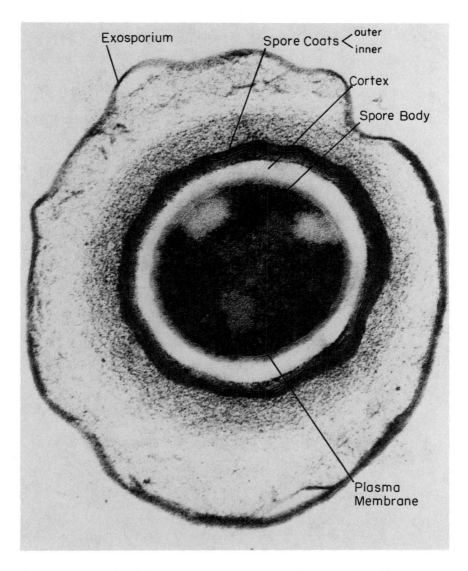

Figure 3-4 Electron micrograph of a thin section of a spore of *Bacillus sphaericus*. The cortex is electron-transparent owing to the particular fixation procedure that was used. The magnification is 158,534×. (Courtesy of Dr. S. C. Holt)

4. Spore Cortex

The spore cortex is bounded by the germ cell wall on one side and by a membrane derived from the outer forespore membrane (OFsM) on the other. It consists solely of peptidoglycan but is substantially modified from that found in the germ cell wall or in the vegetative cell wall. Only 30% of the muramic acid residues are substituted with the full tetrapeptides. Of the remaining muramic acid residues, 15% are substituted with a C-terminal L-alanine residue while the remaining 55% exist as the lactam of muramic acid (Figure 3-5). Such a peptidoglycan has physical properties substantially different from those of a more conventional wall peptidoglycan. It is gel-like, far less resistant to stretching or shrinking and, because of the muramic lactam, much less susceptible to the action of lysozyme. The function of this modified peptidoglycan is not at all clear but will be discussed further in the section on spore resistance.

5. Spore Coat

The spore coat[7] is bounded on its inner surface by what was originally the OFsM, and on its outer surface by the exosporium. The spore coat is defined morphologically and operationally. In *Bacillus cereus*, it is a three-layered structure that when examined by freeze-etch techniques was shown to comprise an outer layer that is cross-patched (CP), an underlying pitted (P) layer, and an innermost undercoat (UC). Operationally, it is what remains of the spore after it has been broken, exhaustively washed, and treated with lysozyme. The spore coat contains 30 to 60% of the spore dry weight, 80% of the spore protein, and may occupy 50% of the spore volume. In *B. cereus*, the spore coat composition has been characterized extensively and consists of a single polypeptide with a molecular weight of about 12 000. The protein is exceptionally rich in cysteine, which may permit extensive tertiary structure based on disulfide bridges, hydrophobic interaction, and ion pairs. Thus, the morphological complexity of the spore coat of *B. cereus* may be a reflection of relatively minor modifications of a single polypeptide; as such it offers the possibility of describing morphological configurations in terms of molecular configuration. (It should be pointed out that a somewhat different situation exists in *B. subtilis* where the spore coat seems to be composed of 14 different species of polypeptide.)

The spore coat is extremely resistant to a variety of drastic chemical treatments. This fact suggests that the coat may play a role in spore resistance. However, mutants severely deficient in spore coat composition and organization retain their heat resistance. Thus, while the coat may have other functions with regard to spore resistance, it clearly plays no role in heat resistance.

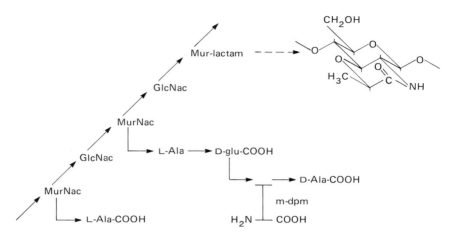

Figure 3-5 Structure of peptidoglycan from the cortex of spores of *Bacillus subtilis*. (Adapted from Tipper et al. 1977)

6. Exposporium

The *exosporium* is the outermost spore layer and its relationship to the spore varies from species to species. In species such as *B. cereus*, it is a thin, loose-fitting structure that can be removed easily by mechanical shear. In other species such as *B. subtilis* or *B. megaterium*, it is more closely fitting and may in fact have some close physical connection with the spore coat. The exosporium of *B. cereus* has been chemically characterized and consists of 52% protein, 20% polysaccharide (glucose : rhamnose : glucosamine : ribose; 13 : 9 : 15 : 1), 12.5% neutral lipid, 5.5% phospholipid (cardiolipin), and 3.8% ash. The protein composition of the exosporium is distinguished from that of the outer coat by a low content of cysteine. Furthermore, drastic solubilization of the exosporium reveals eight bands of protein upon gel electrophoresis. The exosporium proteins are highly resistant to proteolytic enzymes, which suggests (but of course does not prove) that it may play a role as the outermost defense barrier of the spore.

D. The Life Cycle of Bacillus

The life cycle of *Bacillus* illustrates a point made in the Introduction, that this organism has an alternative to continual growth. When the environment is no longer favorable for growth, an equilibrium between growth and sporulation is shifted so that sporulation now becomes the dominant mode. Part of each individual cell is converted to a spore, which is a metabolically dormant, optically refractile, highly resistant cell. The remainder of the cell participates in the sporulation process but eventually lyses, leaving a free spore. This spore will remain as such until conditions for germination are appropriate, at which time it will germinate, becoming once again a vegetative cell.

1. Spore Formation

The sequence of morphological events is illustrated diagrammatically in Figure 3-6. These events are divided into seven stages somewhat arbitrarily, beginning with the first visible change, the reorganization of the nuclear material into an elongated axial filament, and ending with stage VII, the completed spore. (Stage I is now excluded as a bona fide stage of sporulation since it has not been possible to obtain any mutants blocked at the stage of axial filament elongation.) The stages in-between are marked by the partitioning off of the spore DNA by the cell membrane (stage II), the engulfment of the forespore by an inner and outer forespore membrane (stage III), the deposition of the thick, spore cortex (stage IV), the biosynthesis and assembly of the spore coat proteins (stage V), and the completion of cortex and coat synthesis and organization (stage VI). In *B. subtilis*, this process occupies about seven hours.

One of the most unusual features of the process involves the reversal of the polarity of the outer forespore membrane (OFsM);[8] this process is illustrated in Figure 3-7. As you can see, the OFsM retains its original inner-outer polarity; the polarity of the inner forespore membrane (IFsM) on the other hand, is reversed. This unusual morphogenetic result was detected by examination of the sequence of morphological events during sporulation; it was demonstrated by examining the ability of the intact developing spore to show certain enzymatic activities known to be present only on the inner surface of the cytoplasmic membrane. The appearance of these activities was correlated with the appropriate morphological changes and could be shown not to be attributable to a general breakdown of the permeability of the forespore membrane. In view of the fact that transport across a membrane is vectorial (i.e., has directionality) and in view of the fact that various enzymes are spatially oriented in the membrane, the reversal of the polarity must have developmental significance and functional consequences for the cell. It is not at all clear what these may be.

2. Spore Germination[9]

There is far less interest and research on the germination of the endospore than on its formation. The reason for this is not at all clear to me. The

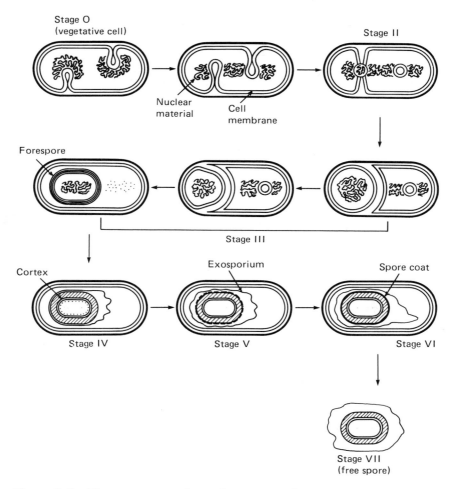

Figure 3-6 The seven stages of sporulation in *Bacillus*.

process whereby spore dormancy is broken, the nature of the dormant state itself, the setting in motion of the complex set of events required for outgrowth—all of these represent a fascinating and relatively unexplored aspect of endospore development.

The process of germination is conventionally divided into three stages: (1) activation, (2) initiation, and (3) outgrowth. Activation refers to the fact that freshly formed spores of *Bacillus* will not readily germinate unless they have been first treated with heat or one of a variety of other treatments (e.g., ionizing radiation, low pH, or reducing agents). Spores that have been allowed to age, that is, have been stored for a period of weeks, will

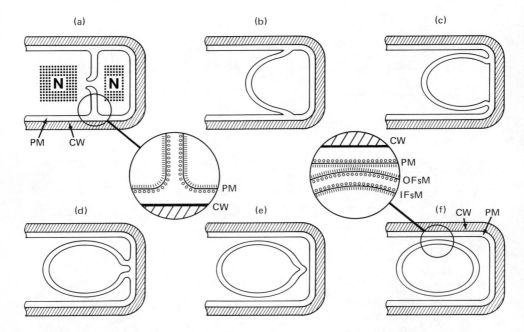

Figure 3-7 Reversal of "sidedness" of the cell membrane during endospore formation in *Bacillus*. N, nuclear material; *pm*, plasma membrane; *cw*, cell wall; *OFsM*, outer forespore membrane, *IFsM*, inner forespore membrane. In the insets, which are expanded views of the membranes and cell wall: *o*, outside; *i*, inside. (Adapted from Wilkenson et al. 1975)

germinate without activation. It is as if the activation accelerates a normal aging process that must precede germination. There are a number of hypotheses to explain the phenomenon, but it remains a puzzle both in terms of its mechanism as well as its biological function.

Initiation is an irreversible process and involves the conversion of the activated spore to a cell that has lost the physiological properties characteristic of the spore. The various physiological events that occur are listed in Table 3-1. Initiation can be effected by a wide variety of treatments including nutritional, enzymatic, chemical, and physical. For example, treatment of activated spores with L-alanine or certain other amino acids, glucose or other sugars, adenine or other bases will initiate germination. Among the nonnutrient chemical initiators are calcium, dipicolinic acid, and some long-chain alkylamines. In addition, treatment of cells with lysozyme or other hydrolytic enzymes or with physical forces such as hydrostatic pressure or abrasion will initiate germination.

In a large sense, the essential function of the initiation phase of germination is to set the stage for the final phase of the process, outgrowth. In other words, during this phase of germination the cell must synthesize those enzymatic and structural proteins necessary for converting a spore to a vegetative cell. This synthesis requires that the metabolically dormant cell begin to do three things: generate ATP, synthesize RNA, and provide amino acids for assembly into protein. The strategy whereby the spore accomplishes this is that the period of macromolecular biosynthesis is preceded by a "turnover" period during which the cell makes use of stored phosphorylated compounds and converts spore RNA and protein to free nucleotides and amino acids. Thus, the pump is primed for the subsequent net synthesis of macromolecules.

The spore does not contain levels of ATP sufficient to allow it to metabolize normally; in fact, its energy charge (a measure of the relative amounts of ATP, ADP, and AMP) is substantially lower than that of corresponding vegetative cells. It does, however, contain considerable amounts of 3-phosphoglyceric acid, which, during initiation, is rapidly converted to pyruvate and ATP. This ATP provides sufficient energy for converting the pool of spore ribonucleotide triphosphates into nucleic acids; the pool of spore ribonucleotides is replenished by the breakdown of about 5% of the spore RNA. Polymerization of these RNA precursors is catalyzed by RNA polymerase that is present in the spore. Amino acid precursors for protein biosynthesis are supplied by the proteolytic breakdown of specific

Table 3-1 Time of occurrence of the various physiological events during initiation of Bacillus endospores

Event	Time[a] for 50% completion of event			
	B. megaterium	B. cereus	B. cereus	B. subtilis
Heat sensitivity	41	54	61	
Chemical sensitivity	41			
Release of K^+				50
Release of CA^{2+}			86	
Release of DPA	64	65	86	
Stainability	77	70		
Phase-darkening	95	60		
Release of hexosamine		92		
Release of DAP			97	
Fall in extinction	100	100	100	100

[a]Time relative to fall in extinction, taken as 100.

Source: From Gould, G. W. and Dring, G. J. 1972. In Spores V (H. O. Halvorson, R. Hanson, and L. L. Campbell, eds.). American Society for Microbiology, Washington, D.C., pp. 401-8.

spore proteins during the first few minutes of initiation. The kinetics of these events are illustrated in Figure 3-8.

Outgrowth, the last stage of germination before vegetative growth, coincides with the onset of net DNA synthesis and is illustrated in Figure 3-9. During this period, the spore is morphologically converted to the vegetative cell, macromolecular synthesis is characteristic of the vegetative cell, and the cell is now poised for vegetative growth.

E. Genetics of Bacillus[10]

Until recently, the genetics of *Bacillus* was limited to the mapping of a set of sporulation mutants. The mapping technique was transformation by naked linear DNA and was inherently limited to fine structure analyses. Nevertheless, a considerable amount of information was accumulated. Within the past decade or so there has been a substantial increase in the

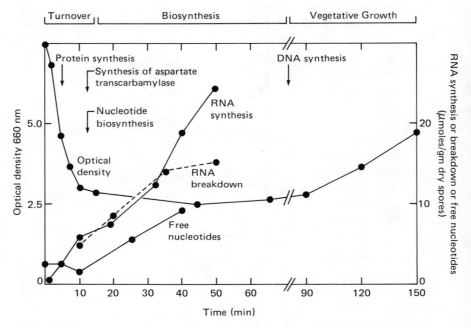

Figure 3-8 Graph showing macromolecular synthesis and nucleotide metabolism during endospore germination in *Bacillus*. (Adapted from Dawes and Hansen 1972)

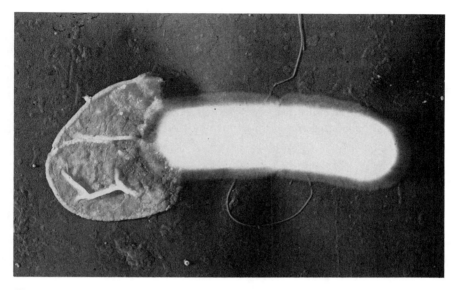

Figure 3-9 Electron micrograph of a gold-palladium shadowed preparation of a germinating cell of *Bacillus cereus*. (From Strange and Hunter 1969)

sophistication and power of the techniques available for genetic analysis in *Bacillus*. Thus, there is now available:

1. Generalized transduction by the very large phage PBS1 that makes possible the rapid mapping of markers, albeit at a coarse level of resolution.
2. Transformation by naked linear DNA that allows fine structure mapping.
3. Specialized transduction with the large temperate phage SPb that makes possible the generation of partial diploids for complementation and dominance analysis.
4. Methods for generating fusions with the lacZ gene so as to facilitate detection of expression of developmental genes.
5. Methods for inserting the streptococcal transposon Tn917 into the genome of *B. subtilis*.
6. Isolation of extragenic suppressor mutations for examining interactions among sporulation proteins and for examining the regulatory aspects of developmental pathways.
7. Methods for isolating and cloning developmental genes of *Bacillus* and for introducing genetically engineered plasmids into the cells.

1. Mutant Studies

Up until recently, the greatest emphasis by far in the study of *Bacillus* genetics has been on the generation and analysis of sporulation mutants. This emphasis has allowed not only the assessment of the developmental function of various physiological and biochemical properties but has also been the starting point for constructing genetic maps for *Bacillus* (Figure 3-10). First I shall describe the various kinds of mutants that have been examined.

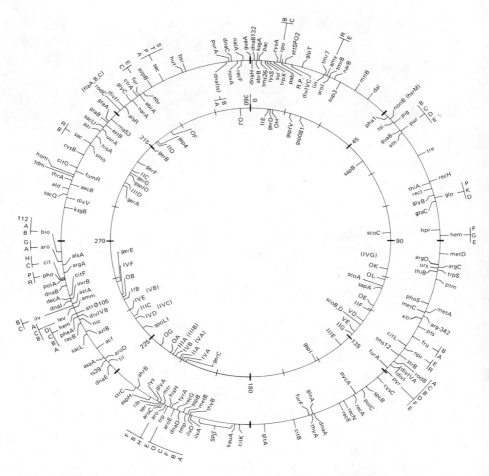

Figure 3-10 Genetic map of the chromosome of *Bacillus subtilis*. For reference purposes, the map is divided into 360° and these numbers are noted on the inner circle. The inner circle contains sporulation and germination loci; all others are on the outer circle. Owing to the lack of space, not all of the known loci are included on the map. (Adapted from Henner and Hoch 1980)

These can be divided on the basis of the kinds of properties that are mutated, that is, developmental, physiological/biochemical, and growth/DNA cycle mutants. A further, useful subdivision is based on the nature of the mutation (i.e., conditional or nonconditional).

a. Developmental mutants. These mutants are classified as asporogenic or oligosporogenic; the former *(spo)* are completely unable to sporulate, and the latter *(osp)* results in a reduced number of heat-resistant spores. Opinion seems to favor the notion that the *osp* mutation is not a unique mutation in a separate locus distinct from an *spo* mutation; rather, the *osp* mutation in most cases seems to be simply a leaky mutation. Thus, the significance of *osp* mutations per se is probably rather trivial. Before proceeding with a description and discussion of mutations in so-called sporulation genes, it is useful to consider the various categories of genes that may appear to be involved in sporulation.

One category is those genes coding for enzymes that function and are synthesized during both vegetative growth and sporulation. These would involve enzymes required for energy generation, macromolecular biosynthesis, and so on. A second category would be genes coding for those enzymes that appear during sporulation because they are induced by the same nutritional circumstances that induce sporulation but do not play any role in the sporulation process. This group includes, for example, enzymes of the arginine breakdown pathway. A third category of genes would be those coding for enzymes required for sporulation, appearing or increasing during sporulation, but not serving any function unique to the spore. An example of such are the enzymes of the Krebs cycle, which appear during sporulation but are absent during normal vegetative growth. The final category is of those genes coding for enzymes unique to and indispensable for sporulation. This category includes the synthesis of spore cortex peptidoglycan, dipicolinic acid, spore coat protein, and certainly others. (This discussion is largely theoretical since, of all of the *spo* mutants that have been isolated, none has been characterized in terms of the nature of the gene product affected.)

Spo mutants may be those blocked either at initiation (referred to as *spo0*) or at some stage during sporulation. (No distinction is made between stages 0 and I. All such mutants are classified as *spo0*.) *Spo* II mutants have sporulation septa, but no forespores; *spo* III mutants contain a forespore but no additional spore structures such as cortex, spore coat, and so forth; *spo* IV mutants have a forespore surrounded by a cell wall and some cortex but no spore coat; *spo* V mutants have cell wall, cortex, and spore coat in varying degrees of completion. There are nine different *spo0* loci, seven *spo* II, five *spo* III, seven *spo* IV, and five *spo* V loci.

There are also mutations that alter the timing of developmental events, mutants that will sporulate in rich media in which sporulation is ordinarily prevented (derepressed mutants), and mutants that produce spores that appear normal but are otherwise deranged (e.g., spores that have lowered heat resistance or impaired germination ability).

b. Physiological/biochemical mutants. A variety of mutants have been isolated on the basis of various physiological or biochemical lesions, that turn out also to be developmentally impaired. These sorts of mutants are especially valuable as they permit one to associate development with a specific, well-defined physiological process. Among the most useful of these sorts of mutants are those isolated on the basis of their resistance to certain antibiotics. For example, some mutants that are resistant to antibiotics that inhibit transcriptional or translational processes have been shown also to be blocked in sporulation. Reflecting another aspect of this approach, there have been reports that mutants of *Bacillus brevis* which do not synthesize the peptide antibiotic gramicidin S produce spores that have a decreased heat resistance and a reduced DPA content. This important finding provided evidence that antibiotics, which are almost invariably produced by spore-forming microbes (e.g., *Bacillus, Streptomyces, Myxococcus, Penicillium*), may play a regulatory role in the sporulation process itself. A recent publication, however, reported the inability to repeat these results, and so the matter is still essentially unresolved.[10A]

c. Growth/DNA replication mutants. While such mutants are obviously a subclass of the group referred to above, there is such an intimate and apparently causal association between growth and DNA replication on the one hand and sporulation on the other that it is worth considering the group separately. Certain mutants that are blocked at the initial stage of sporulation *(spo0)* have been shown to form vegetative cells half the normal length; and mutants temperature sensitive for initiation or completion of a round of DNA replication are also found to be *Spo0* mutants.

d. Conditional mutants. These are mutants whose mutated property is only expressed when the cell is subjected to a restrictive condition. The most common such condition is elevated temperature, (e.g., 42 C compared to the optimal temperature of 30 to 34 C); it may also be reduced temperature (cold-sensitive mutants) or a nutritional deprivation (e.g., the absence of thymine in a thy-requiring mutant). This procedure allows the isolation of mutants whose mutated gene is normally necessary for vegetative growth. In other words, the mutation would otherwise be a lethal one (e.g., a mutation in DNA replication). The examination of such mutants has not only allowed investigators to determine if there is a relationship between development and a variety of physiological processes but also to determine the order in which various genes are expressed. If there is a temporal sequence of expression of genes during development, then one should be able to determine the time at which a particular gene is expressed and how long it must be expressed, by imposing the restrictive condition (e.g., high temperature) at varying times during development. If the gene in question has already been expressed, the restrictive temperature will have no effect on development; if not, development will be inhibited.

Before leaving the subject of mutations, a word of caution is in order. *Spo* mutations are highly pleiotropic; that is, in a linear dependent sequence of events, many of the properties downstream from a mutated gene will be altered. It is necessary to be able to distinguish this from the results of multiple mutations in the cell. Furthermore, in order to correlate a physiological event with a developmental one, it is necessary to be certain that the mutation that caused the physiological event is the same one that resulted in the developmental change. There is a high probability that the use of certain mutagens will induce multiple mutations; in fact, it is even possible that spontaneous mutations may be multiple. Thus, it is always necessary to determine that the mutant strain is isogenic with the parent. The likelihood that this is indeed the case can be greatly improved by backcrossing the mutation into the parent strain via transduction. Even more effective, however, are reversion studies that demonstrate that the two properties in question are simultaneously lost upon back-mutation.

2. Transformation

Transformation is the process whereby free DNA is taken up by a bacterial cell, undergoes recombination with the host cell genome, and results finally in a recombinant genome. The process was first discovered in *Pneumococcus*, and the recognition by Avery and his coworkers in 1941 that the specific transforming substance was in fact DNA could be said to have been the beginning of molecular biology. Transformation has also been described for some species of *Bacillus* and along with studies of transduction has been used for mapping various genes in *Bacillus*. The pieces that can be transferred via transformation are considerably smaller than those that can be transferred by viral transduction; transformation has thus found its principal use in fine-structure mapping. Cells are capable of binding and taking up donor DNA only at certain stages of their growth cycle and only under fairly specific nutritional and environmental conditions. Such cells are said to be "competent" and will bind double-stranded duplex donor DNA. Such DNA is cleaved on the cell surface, degraded to single-stranded DNA with a maximum molecular weight of about $6-7 \times 10^6$ daltons, and taken up by the host cell. There is a synoptic, covalently bonded union of donor fragment with the host DNA resulting in a heteroduplex; subsequent DNA replication of the heteroduplex results in the generation of two dissimilar homoduplexes. One of these is contained, of course, in the recombinant cell.

3. Viral Transduction

Most transductional studies in *Bacillus* that have been used to map *spo* loci have been carried out with the large phage PBS1. This phage can transfer about 5% of the donor chromosome—a piece roughly 20 times larger than the transforming fragments. Transductional analysis has been used to determine the genetic linkage between *spo* mutations and other markers (usually auxotrophic) by measuring cotransduction frequencies. (See Section F of this chapter for a description of the genetically useful phage SPβ).

4. Protoplast Fusion

Bacillus genetics has long been hampered by the lack of an easy and reliable method for studying genetic complementation. Thus, it has not been possible either to determine the number of genes in a locus or to examine dominance relationships of developmental mutations. In 1978, David Hopwood described a method for fusing protoplasts of *Streptomyces* involving polythylene glycol and similar to the methods previously devised for mammalian cells. Fusion joined the cytoplasm and genetic contents of the two cells and resulted in genetic recombination. Investigators had previously shown that in a similar fashion it was possible to fuse protoplasts of *Bacillus*, but there was an important difference that resulted in genetic complementation via temporary diploidy rather than recombination. Some years ago it was shown in a series of ingenious experiments in Patrick Piggot's laboratory that during sporulation in *B. subtilis* the separate genomes of the mother cell and the forespore remained functional during much of sporulation, and successful sporulation required the expression of both genomes.[11] Thus, in a clever extension of these experiments, Dancer at Oxford[12] showed that it was possible to fuse protoplasts from vegetative cells of a wild-type strain with protoplasts from a sporulation mutant. If the mutant protoplast was prepared at the point of forespore engulfment (stage III), fusion with the parent protoplast resulted in complementation (rescue of sporulation). While it was not possible formally and rigorously to exclude the possibility that complementation resulted from recombination rather than from diploidy, the results argue in favor of the latter alternative. The complementation has been shown to be effective with stages III, IV, and V mutants; thus, complementation studies with all these loci become feasible. The cooperative interplay of the mother cell and forespore genomes during sporulation is reminiscent of the intracellular interactions between mitochondrial or chloroplast genomes and the nuclear genome in eukaryotic cells.

5. Plasmid and Recombinant DNA Genetics[13]

The remarkable developments of recombinant DNA technology have revolutionized *Bacillus* genetics. Previously, classical genetics allowed one to ask a rather limited set of questions about *Bacillus* sporulation. In addition to the mutant studies already referred to, one could ask, how many sporulation genes are there? And to some extent, what is the sequence of expression of these genes? It was not possible to do transacting, *cis*-dominant tests, complementation analyses, nor was it possible to identify a particular *spo* mutation with a specific biochemical defect. It is now possible to do this and more. For example, the fact that it is now possible to isolate and clone sporulation genes allows one to ask a set of important and fundamental questions about the properties of the genes themselves: To what extent are these genes differentially transcribed by vegetative and developing cell RNA polymerases? To what extent can such differences be ascribed to developmental alterations in promoter sites, sigma factors, and so on? By inserting plasmids that remain stably autonomous and that contain sporulation genes and, alternatively by means of protoplast fusion, it is now possible to establish stable diploids and thus perform complementation and dominance analyses. It should be possible to do in vitro coupled transcription-translation with isolated sporulation genes and thus to determine the nature of their gene products.

Special emphasis should be placed on the fact that it is possible in *Bacillus* to create genetic fusions between promotorless *lacZ* genes and the regulatory elements of developmental genes. Thus, it is possible, by using a simple and quick chromogenic test to measure the activity of the regulatory unit by measuring the expression of the indicator (e.g., *lacZ* gene). This approach will be discussed more fully in Chapter 12.

Until recently, it was not clear that *Bacillus* genetics had contributed any insights into the nature of the regulatory processes of *Bacillus* development. It is now obvious that it had, in fact, laid the groundwork for an analysis of development that will now make important contributions to understanding the process.

F. Bacteriophage for Bacillus[14]

It is axiomatic among microbiologists that there is, somewhere in nature, an infecting bacteriophage for every species of bacterium. The ecological and evolutionary implications of this are unclear, if not fascinating. In *Bacillus*, phages have been found for most species; where they have not it is not clear that a substantial effort has indeed been made. In addition, in the case of some of the insect pathogens, the inability to grow the host bacterium under conditions optimal for the experimental demonstration of bacteriophage may be responsible. In any case, the overwhelming amount of work on *Bacillus* phages has been done with three species, *B. megaterium, B. cereus,* and *B. subtilis;* and of these, perhaps 90% has been done with *B. subtilis.*

Interest in *Bacillus* phages has sprung from a number of sources; principal among these is the fact that they have been useful as a tool for the genetic and molecular examination of sporulation. Their utility has overshadowed the fact that they are themselves intrinsically interesting. They include PBS1 and SP15, which are among the largest phages known, and ϕ29 and ϕ15, which are the smallest, double-stranded DNA phages. Some contain a number of unusual DNA bases, such as hydroxymethyluracil and hydroxypentyluracil, and one phage contains DNA that is glycosylated. Some will only infect motile strains and others only nonsporulating strains. Physiologically, they include virulent phages, lysogenic phases, generalized and specialized transducing phages, and pseudolysogenic phages. These latter are phages that exist in some peculiar, quasi-temperate relationship with their host. The phage DNA exists free in the host cell in contrast to the integrated state of the true temperate phage. In that sense it is similar to the *E. coli* phage P1. Unlike P1, however, it seems to be continually released, reinfecting the culture.

The large phage PBS1 has been most useful as a transducing phage for gene mapping. Its large size allows it to transfer as much as 5% of the *Bacillus* genome, and many of the *spo* mutations have been thus mapped.

One of the most interesting and useful observations regarding *Bacillus* phage was that it was possible for vegetative cells to be infected during the early stages of sporulation, and the resulting spores contained phage entrapped at some incomplete stage of viral development. The vigorous pursuit of this phenomenon with *B. subtilis* and phage ϕe led to the finding that, during sporulation, there was a change in the transcriptional specificity of the cell RNA polymerase, so that it was no longer able to transcribe the phage DNA. This result was presumably a consequence of the shift from a vegetative cell specificity to a spore specificity. It was subsequently shown that the shift was a result of a change in the nature of sigma or sigmalike factors associated with the RNA polymerase. This finding was most exciting as it offered for the first time a definable biochemical change that seemed causally involved in sporulation. Now that it has been possible to clone certain sporulation genes it is now feasible to examine the transcriptional specifications of the modified developmental polymerase, using a specific sporulation gene as a template.

Recently it has been possible to infect *B. subtilis* with the large temperate phage SPβ.[15] This phage can lysogenize a host cell even though that cell lacks the chromosomal attachment sites that normally serve as the locations for insertion of the phage chromosome. Thus, the phage can insert at any one of a number of secondary chromosomal sites. Upon induction of phage growth the phage picks up a piece of neighboring host chromosome and acts as a specialized transducing phage. In addition, there is a derivative of SPβ that contains the streptococcal transposon Tn917. This transposon can insert at a number of sites on the host chromosome and generates a mutation in the gene into which it has been inserted. The transposon carries a gene for antibiotic resistance and represents a convenient, selectable marker close to the site of its insertion. Finally, the large size of the SPβ genome (126 kilobases) allows it to accumulate as much as 10 kilobases of cloned DNA without affecting the ability of the phage to replicate. Thus, it allows the introduction into cells of single copies of cloned DNA with relatively high efficiency. The versatility of this sort of manipulation will be discussed in Chapter 12.

G. Induction of Sporulation

During discussions of the transition between growth and sporulation, one frequently encounters the notion of a trigger of development. This concept implies that there exists a sharply focused point at which the cell is switched from one mode to the other. The following illustration may help to point out the inappropriateness of this idea. Imagine a double-beam balance with a weight on one pan. Weights are placed on the opposing pan with no visible effect on the balance pointer. However, at some point the addition of one more weight causes the pointer to move abruptly across the dividing line. An observer who wandered onto the scene just at that point might conclude that the placing of the last weight was the trigger that caused the movement of the pointer. An observer present throughout the process might have more difficulty in determining just what, in fact, the trigger was.

For the purpose of this discussion, it is useful to divide the process of induction of sporulation into two parts: first, those events that collectively set into motion the shift to sporulation; and, second, the point or points of no return or commitment—the points at which the various events leading to sporulation can no longer be reversed.

It is clear that the events that set the process of sporulation in motion are nutritional ones. The earliest direct observation bearing on this was that if vegetative cells of *Bacillus mycoides* (now referred to as *B. cereus*) were placed in a buffered salt solution the cells sporulated; the process could be prevented if glucose was added to the suspension.[16] Subsequently, it has been shown, with many species of *Bacillus* that limitations of carbon, nitrogen, and, in some cases, phosphate will induce cells to sporulate.[17]

Furthermore, this "step-down" stimulus must be applied to the cells during a brief and well-defined period of time during chromosome replication. Using synchronously growing cells of *B. subtilis*, it has been shown that the step-down stimulus must be applied 15 minutes after the initiation of chromosome replication; if the cells are placed in the induction medium after that time, they must remain in it until the next round of chromosome replication has taken place.[18]

There are two general models to explain the process of initiation of sporulation. One proposes a factor or factors produced during vegetative growth that inhibit sporulation and/or a factor or factors that turn on sporulation.[19] The second model is more vague and subtle and proposes that sporulation, like growth, is the result of a delicate and intricate balance of the synthesis of many kinds of macromolecules and that these biosynthetic rates are responsive to the availability of substrates to the cell.[20]

The idea that sporulation is initiated by a factor or factors emerged as a result of the analogy between sporulation and catabolite repression in *E. coli*[21] where the idea of cAMP as a positive effector of the lac operon in *E. coli* fits well with the observed role of glucose as an inhibitor of sporulation. Unfortunately, it has not been possible to obtain experimental support for this particular model. For example, members of the genus *Bacillus* do not contain cAMP. On the other hand, Freese's laboratory has obtained convincing evidence that shows that sporulation in *B. subtilis* can be induced by interfering with the biosynthesis of guanine nucleotides.[22] The problem is discussed in more detail in Section B of Chapter 8.

The alternative to the trigger hypothesis is one that has been referred to as the *multiple-step hypothesis*. Rather than postulating an "on-off" switch that controls sporulation, it suggests that sporulation is the consequence of a whole series of biosynthetic shifts that are themselves the response to changing patterns of substrate availability. Thus, the regulatory enzymatic changes that occur are for the most part not peculiar to sporulation but can also be induced by the appropriate nutritional shift. For example, the TCA cycle enzymes that appear during sporulation of *B. subtilis* are also induced by growth on aspartate, malate, pyruvate, and other organic or amino acids.

It seems clear that the multiple-step hypothesis has played an important dialectical role in the debate on the issue of induction of sporulation. It has corrected the rather oversimplified view that led investigators to seek single triggers of the process. However, it too ignores the fact that new enzymes are in fact synthesized, new macromolecules constructed, and regulatory processes that seem to be unique to sporulation are involved. Thus, the truth lies somewhere in between (as is often the case); induction of sporulation is undoubtedly not an abrupt transition between two modes but is rather the result of a whole series of interlocking shifts in the balance of biosynthetic events. In addition, at some point, unique developmental regulatory events undoubtedly come into play.

It is perhaps appropriate at this point to discuss the matter of commitment, for it may be that what has been called commitment appears at the point where the unique regulatory events emerge. Earlier work from Jackson Foster's laboratory had shown that cells of *B. mycoides* (*B. cereus*) would sporulate within 11 hours if suspended in distilled water.[23] This process was subsequently referred to as "endotrophic sporulation" and illustrated the important fact that sporulation, once set in motion, was an endogenous process—that the biosynthetic events involved turnover and conversion of cellular constitutents but was essentially independent of exogenous nutrients. The exception to this was the fact that the process was inhibited by the presence of exogenous glucose. However, glucose was an effective inhibitor only until about 5 hours, long before visible spores had appeared. Thus, it seemed obvious that there was a point of commitment, a point of no return after which events leading to sporulation could not be reversed. Subsequently, it was made clear that there were a number of points of commitment each referring to a different physiological event during sporulation and occurring at a different time.[24] What one sees as the point of no return of sporulation per se is probably only the accumulated irreversibility of a series of individual processes. The extent to which these processes represent unique developmental regulatory events is not yet clear.

H. Regulation of Sporulation

It seems clear that the orderly appearance of events associated with sporulation are, to some extent, a result of the regulated expression of the genes that control these events. There are a number of lines of evidence that lead to this conclusion.[25] Analysis of developmental mutants has revealed that there are at least 45 separate sporulation loci distributed among the first five stages of sporulation; each locus is genetically distinct and may contain a single sporulation gene or a contiguous cluster of such genes representing a sporulation operon. Various mutants are blocked at different stages of sporulation and, when development is induced, will accumulate at that blocked stage.

Furthermore, if by means of DNA-RNA hybridization one examines the nature of the mRNA synthesized during sporulation, it can be determined that new species of mRNA appear sequentially during the first five stages of sporulation. Thus, the first level of gene expression, transcription, seems to play a regulatory role. Finally, analysis of the proteins synthesized during sporulation reveals that proteins appear and disappear during sporulation. The conclusion that the orderly and regulated firing of sporulation genes is playing a key role during sporulation seems inescapable. While a discussion of the regulation of sporulation inevitably focuses on mechanisms of gene expression, it should be emphasized that other levels of control (e.g., substrate availability or enzyme activity) must be involved.[26]

The metabolic reactions that occur during sporulation have been classified into four categories[27] based on the nature of the role those reactions play in the process. These categories have been discussed earlier but are worth repeating. First, there are those enzymes whose function is identical during growth and sporulation and that persist essentially unchanged during development. This group includes such enzymes as those involved in many of the steps in amino acid, nucleotide, and protein synthesis, ATP generation, and so on. The second group includes those enzymes whose activities or biosynthesis change during development but which play no role in the process. This apparently paradoxical situation reflects the fact that the regulation of certain enzymes (e.g., the arginine catabolism pathway and amylase) responds to the same environmental signals as does sporulation. The third category is of those enzymes whose activity or biosynthesis changes during sporulation, are required for optimal sporulation but are not uniquely sporulation enzymes. An excellent example of this category are the enzymes of the TCA cycle, which appear during sporulation but which can also be induced by growing the vegetative cells on acetate. The fourth category, and the one that is usually the most interesting to developmental biologists, is of those enzymes that appear only during sporulation and for which a sporulation function is known. This category would include those enzymes involved in the synthesis of unique spore components such as the cortical peptidoglycan, dipicolinic acid, and spore coat protein.

Obviously, each of these categories suggests a different regulatory mechanism; and the value of such a classification is that it forces one to acknowledge that developmental regulation most certainly involves a complex interplay among a variety of control processes.

1. Control of Transcription

There are essentially three strong lines of evidence that suggest that transcription of sporulation genes is regulated during sporulation. The first of these emerged from the observation that RNA polymerase from sporulating cells of B. subtilis would not transcribe the DNA of Bacillus phage ϕe, despite the fact that polymerase from vegetative cells did so quite efficiently.[28] It was soon shown that this was not a result of any substantial change in the core polymerase but rather in the accessory factors responsible for polymerase recognition of and binding to the appropriate promoters. The ability of sigma factor (σ) to bind to core polymerase was substantially reduced and new factors S^1 and S^2 appeared during sporulation. The loss of binding ability by sporulating cells and the consequent change in the transcriptional specificity of the polymerase could be prevented by treating the cells during sporulation with chloramphenicol, an inhibitor of protein synthesis. The most straightforward interpretation of these results is that there is an unstable antisigma factor made during sporulation and that its formation requires protein synthesis. The inhibition of σ activity substantially reduces the transcription of vegetative genes, and the subsequent appearance of new accessory factors allows the continued transcription of sporulation genes. The explicit test of this model is now available thanks to recombinant DNA technology. Sporulation genes have been isolated and cloned, and it is now possible to use them as a defined substrate to test the transcriptional specificity of sporulation and vegetative polymerases[29] (see Chapter 10).

The second line of evidence is that new species of mRNA, detected by hybridization-competition experiments, have been shown to appear during sporulation.[30]

Third, a rather ingenious strategy, based on the isolation of antibiotic-resistant mutants, has shown that the nature of the RNA polymerase itself plays a changing role during sporulation. The strategy of the experiment is simple. For example, the antibiotic rifampicin inhibits RNA synthesis by binding to the RNA polymerase. Most rifampicin-resistant mutants contain a modified RNA polymerase that, while unchanged in its ability to transcribe vegetative genes, can no longer efficiently bind rifampicin. A small subclass of these mutants is also asporogenous, indicating that the polymerase modification has also prevented the proper transcription of sporulation genes. In an elegant variation of this experiment, a series of rifampicin-resistant mutants were isolated that were temperature sensitive for sporulation and were blocked at different but specific stages of sporulation, suggesting that the polymerase has a number of specificities, each manifested at a different developmental period.[31]

2. Control of Translation

Translational control could occur at a number of different sites, viz. stabilization of mRNA or modification of ribosomes, tRNA or any of the various factors or enzymes required for translation. While there is evidence that all of these changes do in fact occur, many of the experiments are subject to alternative explanations. There are two principal problems: The first is the presence in *Bacillus* especially during sporulation of extremely active proteolytic enzymes that may create artifactual macromolecular changes in ribosomes, enzymes, protein factors, and so on. Where this has not been carefully attended to, irrelevant changes have been described as sporulation-associated modifications.[32] Second, the transition from growth to sporulation is also a nutritional/physiological transition from the exponential to the stationary phase of the growth cycle. Thus, it is not at all clear that the changes attributed to sporulation are not in fact more characteristic of a shift-down. In any case, there is some evidence that translational control does indeed take place during sporulation. The two most critical and convincing elements of the argument are: (1) It has been possible to isolate a large number of single-site mutants of *B. subtilis* resistant to the antibiotic erythromycin. This antibiotic binds to the L17 protein of the 50S ribosomal subunit, and the mutants contain altered L17 protein. Every one of several hundred mutants is temperature-sensitive for sporulation; 100% of the mutants contransform and cotransduce resistance and asporogeny; and revertants to normal sporulation simultaneously regain erythromycin sensitivity.[33] The strategy behind this experiment is the same as that just described for the rifampicin resistant mutants. (2) In vitro translation systems prepared from *B. subtilis* and using bacteriophage SPO-1 RNA as template revealed that initiation factors (IF) prepared from sporulating cells were ineffective. IF from vegetative cells or from stage 0 mutants were effective and the loss of activity by IF from sporulating cells was directed only against SPO-1 RNA.

There are obviously many loose threads hanging from the regulatory tapestry; in fact it appears as an intricate and often tangled complex. However, the broad outlines have emerged; and the increasing sophistication with which in vitro biosynthetic systems can be investigated along with the immense power of recombinant DNA technology has generated a real sense of optimism.

I. Spore Resistance[34]

It is the dramatic heat resistance of spores that is responsible for so much attention having been focused on these extraordinary cells. And it is a bit ironic that, from a biological point of view, the extreme heat resistance of a spore is probably a trivial feature of its overall ability to withstand environmental extremes. Whereas it is likely that soil organisms may have to withstand desiccation and temperature extremes from −40 to 60 C, it is unlikely that the ability to resist boiling is going to be of much advantage. On the other hand, that ability may simply be an overshoot—an inadvertent consequence of the evolution of the resistance mechanism. In other words, it may not be biologically practical to design an organism whose heat resistance stops at 60 C. In any case, what is the mechanism of this extraordinary resistance? The matter is by no means settled; however, there seems to be rather general agreement that resistance is not the result of the intrinsic heat resistance of spore components but rather of the physical environment that surrounds the essential spore macromolecules. There has long been an implicit assumption that calcium dipicolinate played a role in spore heat resistance. A number of mutants that were DPA⁻ have been obtained and, at least in the case of B. subtilis 168, found to have retained its heat resistance. However, the mutant is not fully depleted of DPA, and it almost surely has multiple genetic lesions. The DPA⁻ mutant of B. cereus is not fully heat resistant. Thus, the situation with regard to the role of DPA has not yet been clarified by work with the mutants.

There are two theories that seem at the present to have the most currency. One of these presupposes that spore resistance is based on a lowered content of free water in the spore and that this is regulated by the physical state of the cortical peptidoglycan. It proposes that the dimensions of the cortical peptidoglycan respond to the ionic environment—that it expands and contracts depending on whether or not the negative changes on the peptidoglycan are repelling each other or are neutralized by mobile, cationic counter ions. Cortical expansion thus creates a mechanical pressure that dehydrates the protoplast. The other theory proposes that spore macromolecules are stabilized in a gel formed by chelation of Ca^{++} with dipicolinic acid. While these theories are plausible and are supported by experimental evidence, at the moment it is impossible to choose between them.

Chapter 4

Caulobacter

A. Introduction

Caulobacter was first systematically described by Henrici and Johnson.[1] They obtained it from the freshwater Lake Alexander in Minnesota and recognized it as a member of a larger group of stalked bacteria which they called the "Caulobacteriales." For many years it remained a bacteriological curiosity and an occasional reminder of the fact that not all bacteria were simply rods, cocci, or spirilla. In 1964, Jeanne Poindexter published her now-classic monograph on *Caulobacter*.[2] This work was followed by a series of papers on *Caulobacter* development that served to awaken interest in the developmental biology of *Caulobacter*.[3]

The life cycle of *Caulobacter* is in striking contrast to that of other developing prokaryotes that develop resting stages as an alternative to vegetative growth. In *Caulobacter*, the events of developmental interest are the alternation, during the growth cycle, of a polar stalk and a polar cluster of pili, phage receptor sites, holdfast material, and a flagellum. Thus, the life cycle of this organism exemplifies the idea of development as a series of events strictly ordered in space and time; what distinguishes *Caulobacter* is that those events are embedded in and coordinated with the growth of the cell rather than alternative to it. The emphasis on *Caulobacter* should not obscure the fact that the developmental biology of a related organism, the budding bacterium *Rhodomicrobium*, has recently been the subject of investigation and shows promise as an alternative system with which to ask similar sorts of questions. A virtue of this system is that it manifests not only the polar morphogenesis characteristics of budding bacteria but also goes through a sporulation stage.[4]

B. Life Cycle

Figure 4–1 illustrates the typical *Caulobacter* life cycle. There are two coexisting, functionally distinct cell types in the population, the swarmer cell and the stalked cell (Figure 4–2). Thus, *Caulobacter* also illustrates a true cellular, as well as population, differentiation. The motile swarmer cell, at some point in its developmental cycle, loses its flagellum, which drops off the cell; the cell fashions, in its place, the characteristic stalk. The stalked cell becomes sessile via an adhesive holdfast at the distal end of the stalk. The stalked cell grows vegetatively, repeatedly producing new flagellated siblings. The stalked cell is thus analogous to a stem cell continually producing new swarmers, each of which will swim off and assume its role as a new stem cell. The temporal aspects of *Caulobacter* development involve

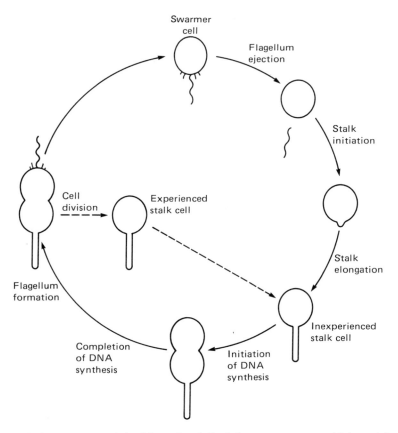

Figure 4-1 Diagram of the life cycle of *Caulobacter crescentus*. (Adapted from Shapiro et al. 1981)

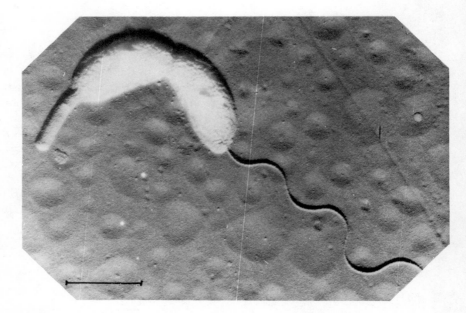

Figure 4-2 Electron micrograph of a shadowed cell of *Caulobacter*. The stalked mother cell is in the process of dividing, resulting in the formation of the flagellated swarmer cell. (From Poindexter 1964)

the timing of the various morphological and physiological events; the spatial aspect involves the polar and coordinated formation of stalk, flagellum, pili, holdfast material, and phage receptors.[5]

C. Ecology and Distribution

The characteristic habitat for *Caulobacter* is an aqueous environment with a low nutrient content. Thus, they can be readily isolated from oligotrophic bodies of freshwater or even tap water. They have also been isolated from sea water and from soil with a high moisture content, neutral to slightly alkaline pH, and a low nutrient content. Their growth is, in fact, inhibited by high nutrient concentrations, and their ecological role seems to be to sweep up the few remaining crumbs of organic matter after such organisms as *Pseudomonas* have finished their feast. It thus becomes possible to rationalize the function of the stalk in terms of adhesiveness and of increased surface area. The ability of the cell to attach via its stalk to a solid surface places it in relatively close contact with a source of organic material.

The high surface/volume ratio of the stalk allows the cell to increase its nutrient uptake capacity by threefold or more. The function of the swarmer cell is probably to allow maximal dispersion of the population. Thus, for an organism whose ecological niche is a low-nutrient, aqueous environment, the *Caulobacter* life cycle seems to represent an optimal and admirable design.

D. Taxonomy and Natural Relationships

There is no familial or ordinal designation for *Caulobacter*, reflecting their uncertain relationships with other Gram-negative bacteria. They are included under the colloquial designation of "budding and/or appendaged bacteria" and share this category with a variety of other bacteria such as *Gallionella, Prosthecomicrobium, Hyphomicrobium, Planctomyces,* and others. There are 10 to 19 species of *Caulobacter*, depending on whom you believe or which set of rules are followed. These species are distinguished on the basis of cell shape, pigmentation, and growth factor requirements. There is a striking lack of information on the biochemistry and physiology of *Caulobacter*. Essentially, all the developmental work has been done with one species, *Caulobacter crescentus*.

E. Cell Structure

The crux of *Caulobacter* development can be encompassed by two connected but separable areas: (1) the temporally/spatially oriented synthesis of the polar, cell-surface structures; and (2) the discontinous processes of cell growth, cell division, and DNA replication. The underlying mechanisms that regulate the second area are not even understood for *Escherichia coli*, where there has been extensive and ingenious experimentation, much less for *Caulobacter*.[6] As a prelude to understanding the regulatory mechanisms that control the time and place of synthesis of flagella, stalk, pili, holdfast, and phage receptors, there has been considerable attention paid to the nature of these structures and to the cell structure generally.

The cell surface of *Caulobacter* is essentially that of a typical Gram-negative bacterium (Figure 4-3). While the ratios of the components that make up the surface of the stalk and the cell are different, the gross composition is the same and includes the components of a typical peptidoglycan. The composition of the outer membrane is unusual in three respects. First,

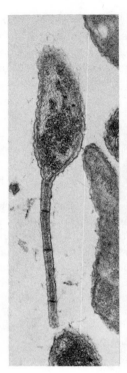
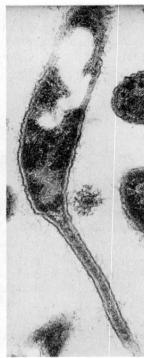

Figure 4-3 Electron micrographs of thin sections of *Caulobacter crescentus* showing the stalk and the annular crossbands. (From Schmidt and Stanier 1966)

the content of ketodesoxyoctanoic acid, a component of the lipopolysaccaride, is present at about 1/10 the amount found in the enteric bacteria. Second, the principal phospholipid is phosphatidylglycerol, rather than the phosphatidylethanolamine characteristic of other Gram-negative bateria. And third, the protein composition is characterized by a predominance of exceptionally large (74 to 130 kilodaltons) proteins. This protein is part of an outer surface array; while such arrays are found on the surface of other Gram-negative bacteria, this one is unusually complex. These observations are interesting, but the inability to derive any developmental significance from them reflects the fact that, in general, one does not fully understand the relationships between chemical structure and function of the cell envelope.

There are a number of interesting aspects of the *Caulobacter* flagellum. One of these is that the flagellum falls off the cell before the development of a stalk at that same site. Neither the mechanisms for the timing of this unusual event nor the mechanism for the actual expulsion is understood. Parenthetically, the subsequent *de novo* synthesis of a flagellum on a new swarmer cell, that has never been flagellated, offers the opportunity to investigate the complete cycle of flagellum formation from the initiation of the basal body through elongation and termination of the filament. An additional interesting feature of the *Caulobacter* flagellum is that the filament, unlike that of most other flagella, is composed of two distinct but immunologically identical flagellin proteins. These are termed flagellin A (26 kilodaltons) and flagellin B (28 kilodaltons) and are coded for by two separate genes. This could reflect either of two possibilities. The first possibility is that the population is heterogeneous, with some cells possessing flagella composed of flagellin A and some of flagellin B. In a sense this could be analogous to the H1, H2 flagellar phase variation that occurs in *Salmonella*. The other possibility, which seems in fact to be supported by preliminary data, is that each flagellum is composed of both proteins. This situation is unusual, albeit not unique; it is difficult, however, to imagine what peculiar function is served by the presence of two flagellins in the filament.[7] The last unusual feature of the *Caulobacter* flagellum is that the basal body whereby the flagellum is attached to the cell contains five successive rings rather than the four characteristic of other Gram-negative bacteria (Figure 4-4).

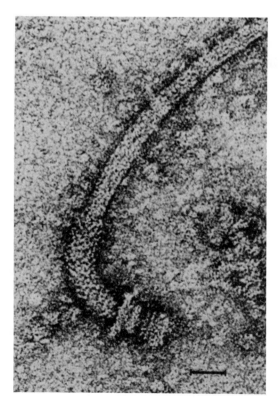

Figure 4-4 A. Electron micrograph of the basal body of the flagellum of *Caulobacter crescentus* stained with uranyl acetate. Note the filament, hook, upper and lower rings, and rod. The bar is 30 nm. (From Johnson et al. 1979)

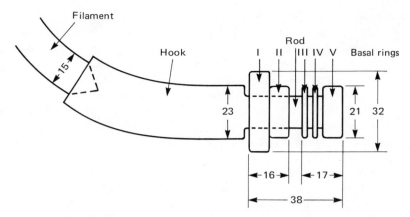

Figure 4-4 B. Schematic diagram of the basal body complex of the flagellum of *Caulobacter crescentus*. Dimensions in nm. (Adapted from Johnson et al. 1979)

Pili, the small, hairlike appendages found on the surface of many Gram-negative bacteria, appear on the predivisional and swarmer cells of *Caulobacter* and then disappear at the same time that the flagellum is shed. Unlike the flagella, the shed pili cannot be found free in the medium; their fate is unknown. They appear at the cell pole, clustered around the base of the flagellum and are composed of a single (8.5 kilodaltons) protein, pilin. Unlike *Caulobacter* flagellins, whose synthesis is stage specific, pilin is made by both stalked and swarmer cells throughout the entire cycle. Regarding the function of the pili, things seem to get stuck to pili as a prelude to a more irreversible, functional attachment; for example, RNA phages and some DNA phages form complexes with the pili before their irreversible binding to the phage receptor sites; and pili seem to participate in cell-to-cell attachment leading to the more stable, holdfast-mediated rosette formation.

Phage receptor sites have been suggested as yet another category of polar, cell-surface structures synthesized by *Caulobacter* in a stage specific fashion. What is not yet clear is whether the polar absorption is a result of the stage-specific presence of polar receptor sites or of the random distribution of receptor sites and the polar concentration of phage adsorption by adhesive pili and holdfasts. Until the nature and distribution of isolated phage receptor sites is characterized, this matter remains unresolved.

The most distinctive morphological feature of *Caulobacter* is, of course, the stalk. It is essentially an extension of the cell surface and consists of cytoplasmic membrane, peptidoglycan, and outer membrane. While the ratio of peptidoglycan components differ, the gross composition of these proteinaceous structures is the same as of those surrounding the cell proper, and a number of envelope-associated enzymes are shared by both. However, enzymes characteristic of the catabolic activities of the cell are absent from the stalk. Isolated stalks are capable of transporting and concentrating glucose with considerable efficiency, but they are unable to catabolize it. These findings are consistent with the view of the stalk as a mechanism for increasing the efficiency of nutrient uptake by the cell. The matter of the shape determination of the stalk is, of course, unexplained. Isolated peptidoglycan sacculi of *Caulobacter* retain the characteristic stalk shape (Figure 4–5); thus, the peptidoglycan may itself be the shape determinant of the stalk. While the gross composition of the *Caulobacter* peptidoglycan is undistinguished, it is likely that the three-dimensional structure

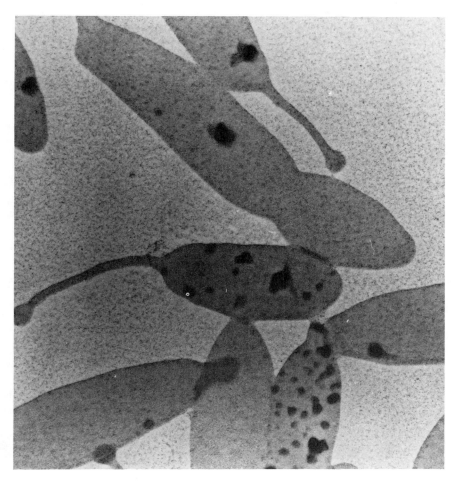

Figure 4–5 Electron micrograph of a stained and shadowed preparation of the peptidoglycan sacculus of *Caulobacter*. (From Poindexter 1981)

of stalk and cell peptidoglycan will, in some unknown fashion, be the shape-determining factor.

Close examination of the stalk reveals the presence of periodic cross-bands along the stalk (Figure 4-6). There is conflicting opinion among the workers in the field as to whether or not these are age-rings, with each marking one division event. In any case, isolated bands appear as donuts rather than as discs. Thus, they seem not to serve to block off one compartment of the stalk from another. However, the openings are small and may

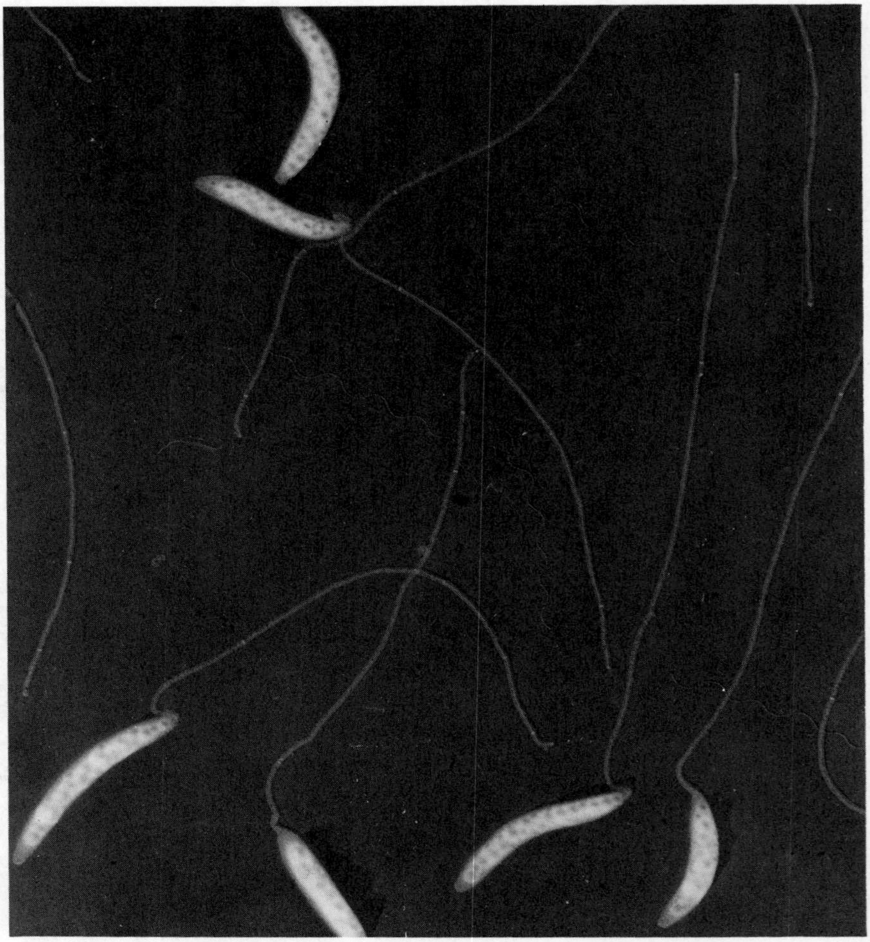

Figure 4-6 Electron micrograph of a shadowed preparation of stalked cells of *Caulobacter crescentus*. Notice the periodic cross-bands along the stalks. (From Poindexter 1981)

effectively retard the passage of cytoplasm, ribosomes, and catabolic enzymes into the stalk while allowing the free diffusion of nutrients into the cell.

F. Physiological Correlates of Development

It is obvious that in order to characterize the regulatory mechanisms that control the various developmental events it would be most useful first to characterize the events themselves. To the extent that one can do so in biochemical terms, one is then in that much better a position to approach the molecular aspects of developmental regulation. The population of *Caulobacter* includes both motile cells, whose function is to swim but not to divide, and stalked cells, which are sessile and continually grow, divide, and produce the motile swarmers. The ability to separate these two cell types by means of differential centrifugation has permitted the correlation of the different phases of cell activity with biosynthetic activity. These are illustrated in Figure 4-7. It is not surprising that DNA synthesis does not take place in the motile, nondividing swarmer cell; and that, conversely, it does in the stalked cell that grows and divides, giving rise to new swarmers.

DNA replication begins immediately upon completion of the swarmer-stalked cell transition or of cell division. It is interesting that the rate of DNA replication appears to vary as some function of the growth rate. Thus, the period of DNA synthesis occupies some constant fraction of the generation time rather than a constant period of time. In this sense it differs from the enteric bacteria.

It should be pointed out that there are two types (or perhaps more accurately two stages) of stalked cells—those that have just shed their flagellum and have newly synthesized a stalk and those that result from cell division. It is not clear whether or not this is a functionally interesting distinction. In any case there is essentially no mechanistic explanation for why DNA synthesis takes place in stalked cells and not in swarmers. Let me emphasize that this does not reflect some peculiar ignorance with regard to *Caulobacter*; the mechanism of regulation of DNA synthesis in *E. coli* is equally obscure.[6]

One step in the process of trying to untangle the regulatory threads involved in the developmental process is to determine the level at which the regulation occurs. Since the stage-specific events one is concerned with in *Caulobacter* involve the assembly of surface structures, one wishes to know whether what is regulated is the synthesis of the constituent macromolecules or their assembly into a structure. Having defined some of the relevant macromolecules (e.g., flagellins or pilin), it is thus useful to ask whether

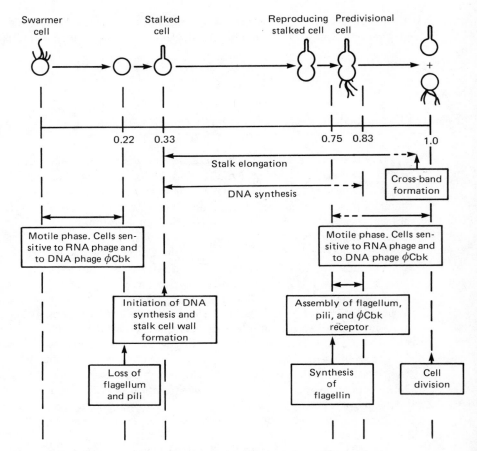

Figure 4-7 Schematic diagram illustrating various developmental and biosynthetic events as a function of the cell cycle of *Caulobacter crescentus*. The numbers are fractions of the complete (1.0) cell cycle. (Adapted from L. Shapiro 1976)

these molecules are synthesized in a stage-specific fashion or, alternatively, throughout the developmental cycle. In a general sense, it has been determined that the relative rates of synthesis of various proteins changed during the developmental cycle. More specifically, flagellin A was found to be synthesized before stalk formation; its synthesis decreased and then accelerated during growth when it was then accompanied by that of flagellin B and hook protein. The synthesis of all of these accelerated during the period of cell division while they were being assembled into flagella. Since *Caulobacter* flagellin, like that of other bacteria, self-assembles into flagella, then there must be a mechanism for sequestering the flagellin macromolecules until the appropriate moment. It is thus interesting that about half of flagellin A is at any moment securely bound to cell membrane. The mechanism and regulatory timing of this process is obscure; likewise, the details of the regulation of flagellin synthesis per se are unclear although the suggestion has been made that since flagellin A synthesis is resistant to rifampicin a relatively stable mRNA is involved.

Pili, like flagella, are assembled in a stage-specific fashion; however, unlike the flagellins, pilins are synthesized throughout the developmental cycle.

G. Caulobacter Phage

There are seven general groups of bacteriophage that infect *Caulobacter*. Six of these are double-stranded DNA phages and are distinguished on the basis of their morphology. The seventh group contains single-stranded RNA phages.[8] There are two aspects of the relevance of the caulophages to development: one is that phage sensitivity is not evenly expressed throughout the cell cycle; phage adsorption seems to occur more effectively to swarmer than stalked cells. Second, transducing phages will, of course, be useful for genetic analysis of development.

The phages adsorb preferentially to either the flagella, pili, or the holdfast end of swarmer cells (Figure 4-8), depending on the host organism.[9] Mutants have been isolated that lack either flagella, pili, or the adhesive holdfast site. Some isolates of each type retain their susceptibility to the respective phages. Thus, it seems that the initial adsorption sites are not

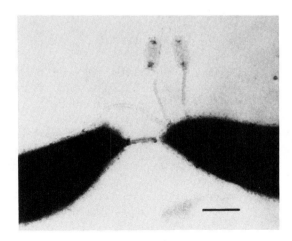

Figure 4-8 Electron micrograph of a stained preparation of *Caulobacter crescentus* showing the polar attachment of phage ɸLC72 to the flagellated end of the swarmer cell. (From Huguenel and Newton 1983)

necessarily the phage receptor sites but may simply enhance the first, reversible association between the phage and its host. Since the phage receptors have not been isolated or characterized, it is not yet possible to determine whether receptor synthesis is, like pili, flagella, holdfast, and stalk synthesis, a stage-specific polar event.

Lysogenic phages have been reported for *Caulobacter*; however, it has not yet been determined if these are capable of specialized transduction. A number of generalized transducing phages have been isolated; phage cr30 is a large phage (MW 10^8 daltons) that mediates generalized transduction at a respectable frequency; the genome is sufficiently large to cotransduce markers separated by as much as 4% of the host genome. A serious inconvenience, however, that compromises the effectiveness of the phage for mapping is that transductional frequencies for a number of different markers varied as much as 400-fold as a function of multiplicity of infection and the nature of the markers chosen. Thus, it seems that the potential value of the system for a primary mapping of developmental genes is not yet fully realized.

H. Caulobacter Genetics[10]

It is now possible to subject *Caulobacter* to genetic analysis making use of a number of sophisticated techniques. In addition to routine generation of mutants, it is possible to move *Caulobacter* genes around by means of conjugal mating and generalized transduction. In addition, plasmids (e.g., RP4) can be transferred to and maintained in *C. crescentus*. Finally, transposon mutagenesis using Tn5 and Tn7 is now possible.

1. Conjugal Mating

Cells of *C. crescentus* have been shown to exchange genetic material by conjugal transfer in a fashion somewhat analogous to the classical conjugation in *E. coli*. However, some important differences exist, the major one being that the populations do not seem to be stably differentiated into male (or donor) or female (or recipient) partners. While the mating is polar, in that DNA is unilaterally transferred from donor to recipient, both can serve either function and have been described by one author as prokaryotic hermaphrodites. There are strains that are more fertile than others—the frequency of gene transfer ranging from about 10^{-7} to 10^{-2}—and while fertile donors have been generated by UV light mutagenesis of nonfertile strains, nothing equivalent to the F-factor of *E. coli* has been described. This conjugation system has not yet been proven useful for genetic analysis.

2. Plasmid Transfer, Chromosome Mobilization, and Transposon Mutagenesis.

The promiscuous plasmids RP1 and RP4 that seem to be native to *Pseudomonas* can be transferred from *Pseudomonas* or *E. coli* to *Caulobacter* and thence from *Caulobacter* to *Caulobacter*. The various drug resistance markers associated with the plasmids are, of course, also transferred, and the frequency of such transfer approaches 100%. In addition, there is some indication of low efficiency chromosomal mobilization allowing conversion of auxotrophs to prototrophs. When these occur between different species, the mobilized genes are unstable; when transfer is intraspecific, the mobilized genes seem quite stable. Use of this system does offer the option of addressing questions of gene dosage and dominance effects; the latter could be most useful for distinguishing between gene effects that are positional versus those that involved the production of a diffusible factor.

Both Tn5 and Tn7 can be introduced into *C. crescentus* via the plasmid RP4. As is the case in other systems, Tn5 exhibits no obvious site specificity; Tn7, however, does insert at a single site. Transductional analysis in combination with three-, four-, and five-factor crosses mediated by the plasmid RP4 have been used to construct a preliminary genetic map for *C. crescentus*[10] (Figure 4-9).

3. Mutants

There seem to be no peculiar or systematic problems associated with obtaining mutants in *Caulobacter*. Thus, mutants involving both selectable and nonselectable properties have been isolated. These include auxotrophy, development, phage resistance, drug resistance, carbohydrate metabolism, cell division, and DNA replication. These latter two groups have been most useful in examining the linear dependence of such events as stalk formation, flagella synthesis, and so forth.

4. Transcriptional and Translational Controls of Caulobacter

Although it seems intuitively reasonable that transcriptional and translational controls must play a role in regulating the developmental events in *Caulobacter*, no such roles have yet been defined. The RNA polymerase is an orthodox, four-subunit structure containing α, β, β'', and σ subunits. When polymerases from different stages of development were examined for their ability to transcribe phage T2 DNA, no differences were found. However, before one reaches any hasty conclusions, it must be emphatically pointed out that this is a test too crude by an order of magnitude to detect

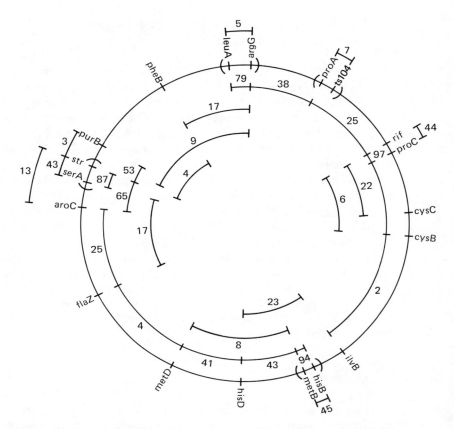

Figure 4-9 Genetic map of *Caulobacter crescentus*. The numbers are to compare conjugation and transduction for measuring the linkage between markers. Numbers within the circle represent percentage cotransfer of markers measured by conjugation; the other numbers represent percentage cotransfer measured by transduction. (Adapted fdrom Shapiro et al. 1981)

the sorts of developmental controls that may in fact exist. Recent experiments indicate quite clearly that the interplay between polymerase attachment to a promoter and subsequent transcription is exquisitely responsive to signals transmitted from the rest of the transcribed unit. Thus, whether or not a particular polymerase will really transcribe a particular gene depends on this interplay, and the test cannot be carried out with heterologous DNA like that from T2 phage or with synthetic polynucleotides as the templates. The use of cloned fragments of developmental and nondevelopmental genes as templates for polymerase from different cell stages offers the direct approach to this question.

With regard to controls at a translational level, it is possible to construct in vitro systems using *Caulobacter* ribosomes and *Caulobacter* phage

RNA as the template. Whereas there is some crude specificity (i.e., the *Caulobacter* ribosomes will not bind RNA from the coliphage MS2), the same caution voiced earlier for examining transcriptional specificity must be emphasized. Until explicit messages that are characteristic of the different stages are available, the question of the presence or nature of translational controls must be held in reserve.

I. Flagella Synthesis

The regulation of the synthesis and assembly of the *Caulobacter* flagella is emerging as the model system of choice for studying *Caulobacter* development. As such it has numerous virtues: (1) The developmental event can be characterized in the precisely defined terms of a set of well-characterized proteins. (2) The synthesis and assembly are regulated in time, as a function of the cell cycle. (3) The assembly is polar and thus manifests the spatial differentiation that is of interest. (4) Mutants are readily available.

One of the problems of considerable interest to those who study the nature of signal exchange between cells is the mechanism whereby signal reception is transduced into a cellular response. The system that seems most exciting and promising in this sense is the process of protein methylation as a response to chemotaxis in *E. coli*. This process involves the methyl-accepting chemotaxis proteins (MCP), the methylesterase and the demethylases. Thus, it is of special interest and relevant to flagellar synthesis in *C. crescentus* that an MCP system exists in *C. crescentus* and furthermore that its appearance and activity varies as a function of the life cycle.[11] Thus, in swarmer cells it is possible to demonstrate MCP methylation in vivo and in vitro and methylesterase synthesis. In stalked cells there is no MCP methylation in vivo or in vitro, and methylesterase synthesis is very low. In dividing stalked cells there is both active MCP methylation and methylesterase synthesis.

It has been possible to clone the flagellin genes by means of restriction endonuclease mapping to study their chromosomal location. In addition, cloning of the flagellin genes has allowed sequence determinations that revealed that the genes contain consensus sequences for polymerase recognition and transcription interaction similar to those found in *E. coli*. Use of cloned flagellin genes has also allowed the examination of the timing of synthesis of flagellin mRNA. These experiments have shown that transcription is confined to those periods of time when the cells are engaged in active DNA synthesis. The relation between transcription and replication is not obvious.[12]

J. Concluding Remarks and Salient Questions

As pointed out in the Introduction, *Caulobacter* exemplifies a particular kind of developmental process—one where the developmental events rather than being alternative to growth are an integral part of growth and cell division. This should not obscure the fact that *Caulobacter* also exemplifies a true population differentiation as defined in the Introduction. It is the complex interweaving of growth, morphogenesis, and differentiation that makes *Caulobacter* such a fascinating experimental system. Add to that the fact that there are no essential technical obstacles to doing modern biochemistry and molecular biology with *Caulobacter* and one can understand why it has become such a popular experimental model.

The essential questions then are:

1. *Spatial orientation.* What are the clues that allow the cell to distinguish one end from the other; in effect to regulate the polar location of stalk, pili, flagella, and perhaps phage receptor sites? The specific related questions are: What is the mechanism whereby the polar flagella are shed into the medium? The flagellum that is shed includes the filament and the hook but not the basal body. Are the pili also shed or are they instead degraded or retracted? An understanding of the regulation of this process obviously must be preceded by a more detailed understanding of the process itself. For example, little if anything is known about the structure in *Caulobacter* (or for that matter, in any bacterium) involved in the actual insertion of the basal body of the flagellum into the cytoplasmic membrane. How are new structures generated as part of the normal process of flagellum synthesis? What happens to the structure when the flagellum is shed? Why does the stalk never emerge at a new cell pole (a pole just formed by a cell division)? Or put another way, why must stalk synthesis be preceded by a flagellum-pili complex? Or put yet another way, why doesn't a stalk usually form at both ends? Again, the analogy with cell growth and division may be useful. What is the mechanism whereby *E. coli* regulates the number and location of cell septa? Only one is formed per division cycle, and it is equidistant between the two poles. Two models have been proposed, both of which are relevant to the question of stalk orientation. The diffusible inhibitor model proposes an inhibitor of septum synthesis diffusing from the poles. When its concentration drops below a threshold, a septum is formed. Thus, in the case of *Caulobacter*, the stalk itself would be the source of stalk synthesis inhibitor. The second model, or the mosaic model, postulates that cell surface growth occurs in patches and that the interfaces between patches of growth and of nongrowth create a series of discontinuities that generate patterns.

In some undefined way such a pattern could distinguish between stalked and nonstalked ends. This sort of patchy surface growth is not unknown among microorganisms.

One approach to the problem of spatial orientation has been to move the process back one step to the level of orientation of the mRNA that codes for the spatially oriented protein structure rather than to the orientation of the protein itself. For example, it has recently been possible to clone the flagellin genes of *Caulobacter* and to generate a cDNA probe for flagellin mRNA. With this probe, swarmer and stalk cells were examined and the flagellin mRNA was found to be localized in the swarmer cells. The authors concluded that the asymmetric expression of flagellin was mediated through the compartmentation of flagellin mRNA in the swarmer cells.[12]

2. *Temporal regulation.* In addition to being organized in space, the various activities of *Caulobacter* are oriented in time. Pili, holdfasts, flagella, and stalks, in addition to all the normal growth and divisional appurtenances such as septa, are synthesized and assembled only at precisely designated times. At a less structural level, DNA synthesis is also temporally regulated; only stalked cells in the growth phase of the cycle synthesize DNA. On the face of it, that is not surprising; there does not seem to be any advantage for the nondividing swarmer cells to be synthesizing DNA. What is interesting is the mechanism whereby one daughter cell (the stalked cell) begins immediately to synthesize DNA while in the other DNA synthesis is inhibited.

I find it fascinating that many of the essential problems of *Caulobacter* development are the problems of *E. coli* in a more visible and exaggerated form. It would be indeed ironic if the intriguing and, as yet, intractable problems of the regulation of cellular growth and division yielded to an examination of *Caulobacter* rather than *E. coli*.

Chapter 5

Heterocyst Development in Cyanobacteria

A.   Introduction

The cyanobacteria (née blue-green algae) are morphologically and developmentally the most varied and complex of the groups of bacteria.[1] Until recently, modern analytical studies of the biology of cyanobacteria were limited to a relatively few species and emerged from a handful of laboratories. Over the past decade, however, the late Roger Stanier's laboratory at the Pasteur Institute has pioneered the application of classical pure-culture techniques of bacteriology to the cyanobacteria. As a result, a large number of developmentally interesting cyanobacteria spanning a number of major subgroups are now available in pure culture. Thus, for the more adventuresome microbiologist who wishes to beat out a previously untraveled path there is now available an extraordinary spectrum of developmental phenomena including baeocyte formation in the pleurocapsalean cyanobacteria, budding in the chroococcacean cyanobacteria, necridia and trichome breakage in the *Oscillatoria*, and heterocyst and akinete morphogenesis in the filamentous cyanobacteria.[2] Add to all of this the fact that, among the filamentous cyanobacteria, the phenomenon of spacing of heterocysts and akinetes along the filament adds yet another aspect—that of one-dimensional pattern formation. The cyanobacteria are indeed a rich source of experimental material, but since the goal of this chapter shall be to focus on the heterocyst, the field narrows down considerably.

As an interesting aside, it is useful to point out that almost invariably developmental studies in bacteria focus on a single organism, representative of the group or the phenomenon. Thus, much if not most of the work on the mechanism and regulation of endospore development has fixed on *Bacillus subtilis*, in *Caulobacter* it is *C. crescentus*, in myxobacteria, *Myxococcus xanthus*, and among the actinomycetes, *Streptomyces coelicolor* receives the most attention. Likewise, the favorite subject for heterocyst development is the genus *Anabaena*, although work is by no means limited to that organism. The reasons for such focusing of attention on a single organism are obvious and certainly not limited to the developmental biology of bacteria. The more information that accumulates about an organism, the easier it is to put new facts into a rational context. The linear, accumulative quality of science tends to place a premium on extending and widening an existing trail rather than blazing a new one. Despite this proclivity, there is a continual debate among scientists, whether they are young investigators choosing the path of their own research or grant panels deciding whom to fund, as to where the optimal strategy lies.

Among the cyanobacteria, interest in development has centered mainly on morphogenesis and spacing of heterocysts in *Anabaena cylindrica*, and this will form the subject of the present chapter. However, so as to be able to describe those peculiar, differentiated features of the heterocysts, it is first necessary to go over some of the relevant general properties of cyanobacterial vegetative cells.

B. The Vegetative Cell

The cellular organization of a generalized cyanobacterial cell is illustrated in Figure 5-1. There is a typical cytoplasmic membrane, separated by a periplasmic space from a peptidoglycan layer that is usually somewhat thicker than the typical Gram-negative peptidoglycan. This is surrounded by an outer membrane that is basically typical, containing protein and lipopolysaccharide (LPS). However, the LPS in some cases lacks ketodesoxyoctonate, ordinarily typical of LPS. There are often additional outer layers including sheaths, capsules, and in some cases, thin fibrous layers whose function is unknown. The lipophilic photosynthetic pigments (carotenoids and chlorophylls) are contained in flattened, membrane-bound sheaths called *thylakoids* where they are part of the photochemical reaction center and the antenna; the thylakoid layer varies with cultural conditions and from species to species.

The cyanobacteria, along with the *Rhydophyta* (the red algae), contain a unique pigment-protein complex called *phycobiliprotein*. This complex is the major light-gathering pigment of the cell and is contained in an organelle called a *phycobilisome*. These are disc-shaped, are about 40 nm in diameter,

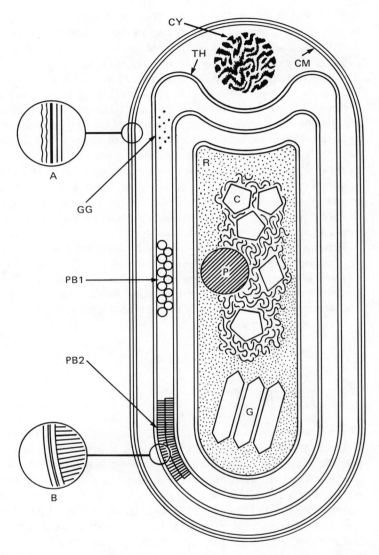

Figure 5-1 Schematic diagram of a cyanobacterial vegetative cell. *CM*, Cell membrane; *TH*, thylakoid; *PB1* and *PB2*, front and side views respectively, of two rows of phycobilisomes, attached to adjacent thylakoids; *GG*, glycogen granules; *P*, polyphosphate granule; *C*, carboxysome, surrounded by nucleoplasm; *G*, gas vesicle. *Inset A*, enlarged view of cell envelope, showing outer membrane and peptidoglycan wall layers and cell membrane. *Inset B*, enlarged view of part of a thylakoid, showing the paired unit membranes with attached phycobilisomes (drawn on one face of the thylakoid only). (Adapted from Stanier and Cohen-Bazire 1977)

and are arranged along the outer surfaces of the thylakoids. The phycobiliprotein may represent 40 to 50% of the total soluble cellular protein of some cyanobacteria.

Another organelle that has significance with regard to heterocyst development is the carboxysome. These organelles, also called "polyhedral bodies," are only found in prokaryotes that fix CO_2 via the reductive pentose phosphate pathway. The carboxysomes are bounded by a thin, nonunit membrane and are packed with ribulose biphosphate carboxylase.

There are three general aspects of cyanobacterial metabolism that bear on heterocyst development: photosynthetic metabolism, nitrogen metabolism, and carbon metabolism.

Cyanobacteria, like green plants, contain two photosystems referred to as I and II. Photosystem I carries out cyclic photophosphorylation, its main light-gathering pigments are chlorophyll and carotenoids, and it is thus generally analogous to the photosynthetic metabolism carried out in all photosynthetic organisms. Photosystem II, on the other hand, generates net reducing power (in collaboration with photosystem I), results in the production of O_2 via the photolytic split of water and participates in what is generally referred to as noncyclic photophosphorylation. Cyanobacterial photosystem II, however, is unusual in that the phycobiliproteins are the principal light-harvesting pigments for photosystem II.

The central feature of the carbon metabolism of cyanobacteria is CO_2 fixation. This takes place via the Calvin cycle and is directly mediated by the incorporation of CO_2 into ribulose biphosphate to give two molecules of phosphoglyceric acid. The reaction is catalyzed by the ribulose biphosphate carboxylase stored in the carboxysomes and also present in soluble form.

Assimilation of fixed nitrogen takes place in vegetative cells by means of a rather straightforward process similar to that which occurs in heterotrophic nitrogen-fixing bacteria. Ammonia is fixed into the amide group of glutamine by means of glutamine synthetase (GS) and then into the amino group of glutamic acid catalyzed by a glutamine-oxoglutarate amido transferase (GOGAT).

This, then, is the life style of the vegetative cyanobacteria. By means of photosystem I, containing chlorophyll and carotenoids, light energy is converted to ATP. In photosystem II, light is gathered by the phycobiliproteins and, with the lipophilic pigments, converted to reducing power, ATP, and oxygen. CO_2 is fixed into phosphoglyceric acid by the ribulose biphosphate carboxylase contained in the carboxysomes; along with glutamate generated by the action of GS and GOGAT, a net influx of biosynthetic starting material is now available to the cell. But what if fixed nitrogen becomes limiting? In what way does the well-known ability of the cyanobacteria to fix dinitrogen enter the story? At this point the stage is set for a discussion of the heterocyst.

C. The Heterocyst[3]

The heterocyst represents a simple and effective solution to what might seem a difficult contradiction. The nitrogenase enzyme that is centrally involved in the fixation of dinitrogen is extremely oxygen sensitive. Furthermore, photosystem II, characteristic of green plants and cyanobacteria, generates O_2 as a result of the photolytic split of water. This seems like a fundamental incompatibility, but the problem is solved by dividing the functions between two specialized cells. Nitrogen fixation takes place in the heterocyst that lacks photosystem II, is thus anoxygenic, and is surrounded by an oxygen-impermeable cell layer; and noncyclic photophosphorylation, generation of O_2 and reductant and CO_2 fixation takes place in the vegetative cells. Solution of one problem, however, creates a spectrum of others. For example, how does the cell optimize the ratio of the two cell types and their physical relation to each other; that is, in a linear filament how do the cells optimize and regulate a one-dimensional spacing problem? How does the cell partition out the various catabolic and biosynthetic functions for maximum safety and efficiency? How does the organism organize optimal mechanisms for cellular interchange? The bare outlines of some of the answers are already available and will be discussed in a later section.

1. Isolation

The heterocyst is more resistant than the vegetative cell to mechanical and chemical stress. The history of heterocyst isolation is a continuing compromise between the need to separate the heterocyst and the vegetative cell as thoroughly as possible in a short period of time and the need to avoid any physical and biochemical damage to the cell. Invariably, one is accomplished at the expense of the other. Recently, a method has been described

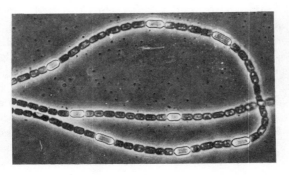

Figure 5-2 Micrograph of a filament of *Anabaena cylindrica* showing the vegetative cells and the interspersed heterocysts. (From Wilcox et al. 1975)

Figure 5-3 Electron micrograph of thin sections of a vegetative cell *(bottom)* and a heterocyst *(top)* of *Anabaena cylindrica*. Note the junction connecting the two cells. (From Lang 1968)

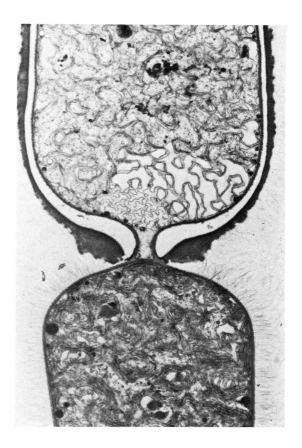

that results in the rapid isolation of what seem to be physically and metabolically intact heterocysts. The method takes advantage of the differential sensitivity of vegetative cells and heterocysts to lysozyme and low-energy sonication and involves the sequential and anaerobic use of these treatments. Using nitrogenase activity and levels of metabolic intermediates as parameters, these heterocysts seem to be appropriate for physiological studies.[4]

2. Structure and Composition

Figure 5-2 is a phase contrast micrograph illustrating the relationship between heterocysts and vegetative cells; the junction between the two is illustrated in greater detail in Figure 5-3. Figure 5-4 is a diagrammatic illustration of the essential features of the heterocyst, some of which may be summarized as follows:

a. Heterocysts are substantially larger than vegetative cells—for example, in *A. cylindrica*, 83 versus 38.7 μm^3.
b. The connection between vegetative cells and heterocysts is a narrow and constricted one whose cross-sectioned area is less than one-tenth that of the connection between vegetative cells. The junction is traversed by a series of fine channels called microplasmodesmata. They

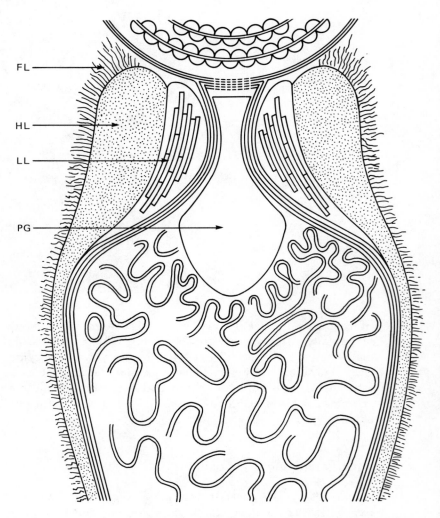

Figure 5-4 Schematic diagram of part of a mature heterocyst showing the polar connection to an adjacent vegetative cell *(top)* through a series of fine pores. PG, polar granule of cyanophycin; FL, fibrous layer; HL, homogeneous layer; LL, laminated layer of the heterocyst envelope. (Adapted from Stanier and Cohen-Bazire 1977)

may play a role in regulating movement of nutrients between the two cells, but, in fact, their function is unknown. In view of the fact that the central feature of heterocyst differentiation involves the biochemical interchange between the two types of the cells, the nature of this junction must be of considerable importance. Little, in fact, is known about the relation between the junction structure and its role, if any, in regulating the metabolic traffic between the nitrogen-fixing heterocyst and the reductant-producing, vegetative cell.

c. At the junction with adjacent vegetative cells, the heterocysts sometimes accumulate a strongly refractile material that has been referred to as polar nodules or granules. These plugs, illustrated in Figure 5-5, have been shown to consist of cyanophycin granules, a storage polymer containing two amino acids, arginine and aspartic acid. The arginine residues are attached to the free carboxyl groups of a polyaspartic acid core.

Figure 5-5 Electron micrograph of a thin section of a heterocyst of *Anabaena cylindrica*, showing the characteristic plug between the heterocyst and vegetative cell. (From Fay and Lang 1971)

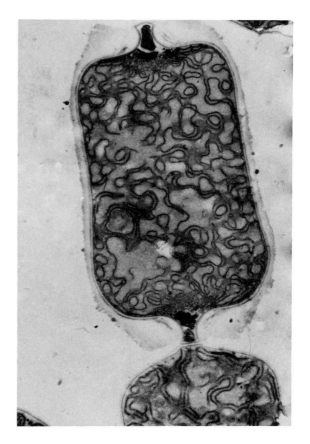

d. The heterocyst lacks the pigment phycocyanin, so that suspensions of heterocysts are yellow-green rather than the characteristic blue-green color. When the culture is subjected to nitrogen starvation, there is a major destruction of the cellular phycobiliprotein, the chromophoric protein contained in the phycobilisomes. When the vegetative cells are fed by the products of the nitrogen fixation of adjacent heterocysts, they resume synthesis of the phycobiliprotein. The heterocysts, however, refrain from resynthesizing the phycobiliprotein. The logic of this strategy is consistent with the role of phycobiliprotein as a light-gathering pigment for photosystem II and the absence of photosystem II from heterocysts.

e. The heterocyst is surrounded by a complex and specialized set of surface layers. These are usually divided into three parts: an outer "fibrous" layer, a central "homogeneous" layer, and an innermost "laminated" layer. The chemical nature of the fibrous layer has not been determined; however, it has been suggested that it may be made up of uncompacted fibers of the polysaccharide that make up the central homogenous layer. This polysaccharide contains as its core repeating subunits of glucose-glucose-glucose-mannose, all connected by β-1,3 linkages. These may have connected to them mono- or disaccharides of glucose, mannose, galactose, or xylose at the 2, 4, or 6 positions. This oligosaccharide network is continuous, except at the poles of the cell where it is penetrated by the microplasmodesmata. The innermost laminated layer is unique, being composed of glycolipids peculiar to cyanobacteria. In *A. cylindrica*, the most abundant such lipid is a glycosylated derivative of the C-26 unbranched alcohol, 1,3,25 hexacosanetriol. It is not unlikely that this layer plays an important role in limiting the permeability of oxygen as well as other solutes into the cell.

f. After many years, the matter of heterocyst DNA seems to be settled. Earlier observations have suggested that the heterocyst lacks DNA or that it is distributed throughout the cell in a diffuse fashion. Chemical determinations have been compromised by the fact that the proximity and close association of heterocysts and vegetative cells hampers the unequivocal separation of the components of the two cell types. In addition, the presence of proheterocysts (developmentally intermediate cells) that may still contain DNA but that have acquired heterocyst resistance have made the data somewhat more difficult to interpret. Thus, most reliance has been on cytological data and these, in fact, have suggested that the heterocyst is anucleate. Current evidence, however, indicates that there is little difference in the complexity or content of DNA between heterocysts and vegetative cells.[5] It does seem, however, that the state of the

DNA or the nature of its distribution in the heterocyst differs from the vegetative cell.

3. Heterocyst Metabolism and Function

As indicated earlier, the essential heterocyst function is to serve as an anaerobic cell within which oxygen-sensitive dinitrogen fixation may occur. In addition, the heterocyst is a dead-end cell; it cannot divide or germinate (as such, it is unique among the prokaryotes). Thus, its biosynthetic activities must be limited strictly to repair and maintenance functions. This section will discuss the heterocysts' anoxygenic photosynthesis with particular reference to pigments and photosystems, dinitrogen fixation and nitrogen metabolism, insulation from oxygen, and biosynthetic limitations.

Originally, the somewhat bleached quality of the heterocysts suggested that they might be deficient in some aspect of their photosynthetic activity. Recall that photosystem II is normally responsible for the generation of O_2 via the photolytic splitting of water; and in the absence of photosystem II the strong reductant and weak oxidant produced by photosystem I recombine, with the resulting phosphorylation of ADP via cyclic photophosphorylation. Recall also that the major light-absorbing pigment involved in photosystem II is the phycobiliprotein. Examination of heterocysts eventually revealed the following:

a. Phycobiliprotein and phycobilisomes had been lost.
b. Hill reaction activity was absent. The Hill reaction occurs in a cell-free extract of photosynthetic plant-type cells and is illustrated by the equation:

$$H_2O + A \xrightarrow{h\nu} \tfrac{1}{2} O_2 + H_2A$$

where A is an appropriate artificial electron acceptor, such as ferricyanide. The Hill reaction is normally present in cyanobacteria.
c. Reduced amounts of chlorophyll, the pigment characteristic of photosystem II; a special photooxidizable chlorophyll referred to as P700 and characteristic of photosystem I is retained in the heterocyst.

In order to fix dinitrogen, the cell must have an abundant supply of energy and of reducing power. The former is certainly ATP, generated in the heterocyst by cyclic photophosphorylation; evidence suggests that the heterocyst may also generate a low potential reductant, which is probably ferredoxin.

There are essentially two aspects of nitrogen metabolism that are relevant to the heterocyst-vegetative cell differentiation. These are the nitrogenase itself, responsible for the conversion of dinitrogen to ammonia and the GS/GOGAT pathway responsible for the transfer of the ammonia to the amino group of an amino acid.

Over a decade ago it was suggested that the ability to fix dinitrogen by cyanobacteria resided solely in the heterocysts.[6] This bold and imaginative proposal was based on less than compelling evidence, and subsequent objections to the idea brought about a sharpening of the data. Improved methods for isolating undamaged heterocysts, increased sophistication of enzymology of dinitrogen fixation, and the elegant use of isotopes has now settled the issue fairly firmly. In cells grown under aerobic conditions, over 90% of the nitrogenase activity of the culture is found in the heterocysts.[7] When cells are grown anaerobically—including the inhibition of photosystem II with dichlorophenyl dimethylurea—both vegetative cells and heterocysts contain nitrogenase activity.[2] Thus it seems clear that the ability to protect nitrogenase from oxygen is limited to the heterocysts and the peculiar function that the heterocyst confers on the organism is expressed only under aerobic conditions.

The fixed nitrogen then finds its way into the main path of nitrogenase biosynthetic precursors via the GS/GOGAT pathway. The enzyme GS, which catalyzes the amidation of glutamate:

$$ATP + glutamate + NH_3 \rightarrow ADP + phosphate + glutamine$$

is present both in heterocysts and vegetative cells but its specific activity in heterocysts is as much as twofold that in vegetative cells.

The second enzyme in the pathway, GOGAT, is present only in vegetative cells where it functions to transfer the amido group of glutamine to an α-keto acceptor:

$$Glutamine + \alpha\text{-ketoglutarate} + NADPH + H^+ \rightarrow + 2\ glutamate + NADP^+$$

Obviously, the net result of all of this is the conversion of one molecule of dinitrogen in the heterocyst to two amino groups in the vegetative cell.

Thus, it is clear that there is a biochemical differentiation between heterocysts and vegetative cells. There seems also to be a structural differentiation that plays a role in dinitrogen fixation; the latter statement has, however, a more conjectural quality about it. The inner, laminated layer of the heterocyst, consisting of the unique glycolipids referred to earlier, seems to act as a physical barrier to prevent oxygen from penetrating the cell. Certain mutants of *Anabaena variabilis* that can fix dinitrogen under microaerobic but not under aerobic conditions are also deficient in the inner layer glycolipids; revertants that regain the ability to fix dinitrogen aerobically also regain the missing glycolipids. This glycolipid layer may also prevent the diffusion of lipophobic materials in general and may thus limit what gets into and out of the cell to the channels between the heterocysts and adjacent vegetative cells.[8]

An important observation was made, based on autoradiographic experiments with $^{14}CO_2$. It was shown that heterocysts failed to incorporate $^{14}CO_2$; however, ^{14}C label from adjacent vegetative cells rapidly diffused into the heterocysts.[9] I wish to discuss this from two directions: first, to rationalize the observation from the point of view of the function of the heterocyst, and second, to comment on the metabolic mechanisms involved. The failure to fix CO_2 makes sense for the heterocyst for two reasons. In view of the fact that the heterocyst is a nonreproductive cell, there is no pressing requirement for net biosynthesis. Furthermore, any CO_2 fixation that would occur in the heterocyst would drain off ATP and reducing power needed for dinitrogen fixation. (It has been estimated that the fixation of one molecule of dinitrogen requires twelve ATPs and six proton/electron pairs).

In autotrophs, CO_2 fixation occurs via the Calvin cycle and the first reaction in the cycle is the fixation of CO_2 into ribulose biphosphate to form two molecules of 3-phosphoglyceric acid. This reaction is catalyzed by ribulose biphosphate carboxylase, and the enzyme is usually contained in polyhedral, cytoplasmic organelles called *carboxysomes*. In *Anabaena*, carboxysomes are present in copious amounts in the vegetative cells but absent in the heterocysts. Likewise, the enzyme itself is absent from the heterocysts as least one other enzyme associated with the Calvin cycle.

With regard to the carbon metabolism in heterocysts there is yet another problem. It is not at all clear what the nature of the reductant for dinitrogen fixation is. There is a tremendous increase in the heterocyst of two of the enzymes of the oxidative pentose pathway. These are glucose-6-phosphate dehydrogenase and 6-phosphogluconate dehydrogenase. The action of these enzymes on glucose-6-phosphate would result in an elevated ratio of NADPH/NADP that in turn could reduce a low potential electron carrier such as ferredoxin. This is, however, conjecture; the actual reductant has not been identified.

The various metabolic exchanges between the heterocyst and vegetative cells are summarized and illustrated in Figure 5-6.

4. Heterocyst Morphogenesis

If a culture of *Anabaena* is allowed to deplete its supply of fixed nitrogen (NH_4^+) or is transferred to a medium free of nitrogen, heterocysts begin to appear along the filament at regularly spaced intervals. Precursor cells, called *proheterocysts,* appear at about 4 to 5 hours, reach a maximum number of about 10% of the total cells in the filament at 16 hours, and begin to be converted to mature heterocysts. These appear at about 14 hours and by 32 hours reach a maximum frequency of 5 to 10%.

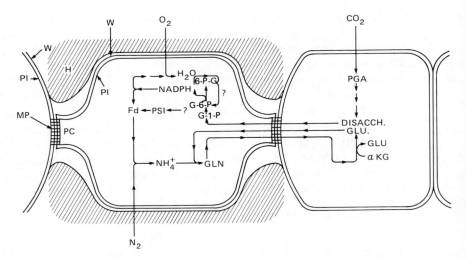

Figure 5.6 Diagram illustrating the principal structural differences and known interactions between a heterocyst *(left)* and a vegetative cell *(right)*. Outside the wall *(W)* of the heterocyst is an envelope consisting principally of a laminated glycolipid layer and a homogeneous, polysaccharide layer *(H)*. Microplasmodesmata *(MP)* join the plasma membranes (Pl) of the two types of cells at the end of the pore channel *(PC)* of the heterocyst. A disaccharide formed by photosynthesis in the vegetative cells moves into heterocysts and may then be metabolized to glucose-6-phosphate and oxidized by the oxidative pentose phosphate pathway. Pyridine nucleotide *(NADPH)* reduced by this pathway can donate electrons to O_2 to maintain reducing conditions within the heterocysts, and can reduce ferredoxin *(Fd)*. Ferredoxin can also be reduced by PSI. Reduced ferredoxin can donate electrons to nitrogenase which reduces N_2 to NH_4^+. Glutamate produced principally by vegetative cells reacts with the NH_4^+ to form glutamine. The glutamine moves into the vegetative cells, where it reacts with α-ketoglutarate to form two molecules of glutamate. (Adapted from Wolk 1979)

Proheterocysts are morphologically and physiologically intermediate between vegetative cells and heterocysts. Depending on their stage of development, they are more or less lysozyme-resistant, they are beginning to synthesize some the the proteins that are characteristic of the heterocysts, but they have not yet synthesized the O_2-impermeable layer found in the mature heterocyst. The process of proheterocyst development is reversible; the presence of a proheterocyst within a putative zone of inhibition of another will result in a competition between the two with one eventually regressing to a vegetative cell.[10] When the spacing observation was originally made,[11] it was suggested that this was a result of a diffusible inhibitor present and evenly distributed in vegetative cells. Upon depletion of fixed nitrogen, the level of inhibitor would fall, a heterocyst would be formed, and it would then serve as a source of inhibitor, generating a one-dimensional gradient along the filament. At the point where the inhibitor concentration drops below a threshold value, a new heterocyst is formed, and so on. In view of the fact that heterocyst development could be inhibited, and, in fact, reversed by the addition of NH_4^+, it was further suggested that the inhibitor was NH_4^+ itself or some product of its immediate metabolism. The main outline of this proposal has retained its currency and, in fact, continues to be a reasonable model. However, it has become clear that neither NH_4^+ nor a direct product of dinitrogen fixation is the inhibitor. Since the proheterocyst appears long before the cell has acquired nitrogenase activity, it is unlikely that a product of nitrogenase activity determines proheterocyst spacing.

Little, if anything, is known at a molecular level about the regulation of heterocyst development. A promising approach that is just beginning to emerge is to focus on the regulation of GS. Since this enzyme is enriched in heterocysts, since the reaction it catalyzes is the only route for ammonia assimilation in the heterocyst, and since its product, glutamine, is an inhibitor of heterocyst development, it obviously must play a key role in heterocyst development. As indicated earlier, unlike GS in other organisms, its activity is not regulated by a reversible adenylation-deadenylation. Nor does the transcription of its mRNA appear to be conventionally regulated. Recently, the gene for GS in *Anabaena* has been cloned in *E. coli*;[12] it is now available as a probe of the transcription program, and we may look forward to experiments that will provide insights into its elusive regulatory mechanisms.

5. Heterocyst Spacing

The examination of the sequence of events that precede the regular spacing of heterocysts along the cyanobacterial filaments has led to a set of descriptive rules:

1. Cell division follows a recursive, asymmetric pattern.
2. If the right- (or left-) sided partner of a cell division is the smaller of the pair, upon its subsequent division the smaller cell will be the right- (or left-) sided member.
3. Only the smaller cells are eligible eventually to become proheterocysts. Thus, with division there is a preliminary restriction of heterocyst eligibility.

These rules are illustrated in Figure 5–7.[13]

The current model proposes then that this restriction plus the zone of influence on either side of a heterocyst combine to control the spacing pattern. The first question is, is the zone generated by a diffusible stimulator of development produced by vegetative cells and destroyed by heterocysts, or conversely, by an inhibitor of heterocyst maturation excreted by

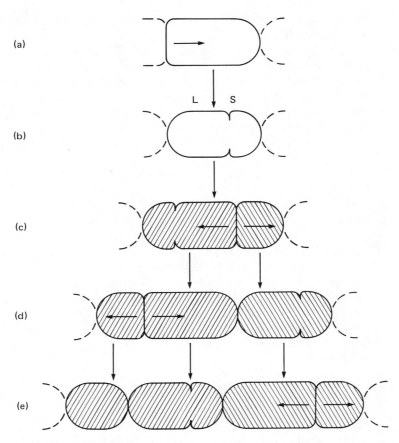

Figure 5-7 Diagram of the division rule for *Anabaena*. *(a)* A cell is represented with an arrow pointing away from the newly formed septum. The rest of the filament is indicated by dotted outlines. *(b)* The two daughter cells are shown. L represents the larger daughter cell; S represents the smaller. Smaller daughters take longer than larger daughters (by some 40%) to reach a subsequent division; this difference introduces an asynchrony which is indicated here. Thus, the division of the first L cell is complete in *(d)* but that of the first S cell only in *(e)*. (Adapted from Wilcox 1975)

proheterocysts?[14] Ingenious microsurgery experiments that result in the generation of filament fragments of varying lengths and heterocyst/ vegetative cell composition fairly clearly point to the latter possibility, namely, the generation by the heterocyst of a developmental inhibitor. Furthermore, it has been suggested that glutamine is a likely candidate for the inhibitor.[4] At this point there is no compelling evidence to support the suggestion; it is, however, consistent with available evidence.

Finally, it is possible to disrupt the inhibition and increase heterocyst frequency either by mutation, by treatment with the amino acid analogs 7-azatryptophan or β-2-thienyl-dl-alanine or with the antibiotic rifampicin. It is interesting that rifampicin, which inhibits the binding of RNA polymerase to its appropriate promoter site, is effective at a concentration two orders of magnitude lower than that necessary to inhibit growth of *Anabaena*.[13]

6. Genetic and Molecular Analyses of Heterocyst Development

Previously, techniques for genetic analysis existed only among the unicellular cyanobacteria, for example *Anacystis* and *Agmenellum*. Recently, however, Wolk's group has engineered a shuttle vector that will replicate in both *E. coli* and in several strains of the filamentous cyanobacterium, *Anabaena*.[15] This vector is based on the *E. coli* plasmid pBR322, contains the cyanobacterial replicon pCV1, and determinants for resistance to chloramphenicol, streptomycin, neomycin, and erythromycin. The key to constructing such a plasmid that is able to replicate in *Anabaena* was the removal from the hybrid plasmid of the sites for *Anabaena* restriction enzymes.

In addition, Haselkorn's group at the University of Chicago has been using recombinant DNA technology to ask questions about the nature of the regulation of expression of nitrogen fixing genes in *Anabaena*.[16] This will be discussed further in Section D of Chapter 10.

Conclusions and Salient Questions

It now seems quite clear that in those cyanobacteria that form heterocysts there is a true population differentiation, with physiological function divided between the two dissimilar cells.

This differentiation allows an organism whose normal photosynthetic activity results in the production of oxygen to fix dinitrogen, a process notoriously sensitive to oxygen. This is accomplished by limiting dinitrogen fixation to the heterocyst, a dead-end cell lacking the oxygen-producing photosystem II, and with substantially reduced permeability to oxygen. These cells are evenly spaced along the filament of vegetative cells. Thus, the developmental issues fall into two categories: the morphogenesis of the heterocyst and the regulation of its spacing. The former category includes the processes that partition the various biochemical activities between the two cell types as well as the biosynthetic processes that control the synthesis and assembly of the various unique heterocyst structures. With regard to the latter category, it is assumed that the spacing is a result of two processes: a recursive, asymmetric cell division and the one-dimensional diffusion of a heterocyst inhibitor. Neither the nature of the asymmetric cell division nor the putative inhibitor is known.

The title of F. E. Fritsch's presidential address to the Linnean Society in London in 1950 was "The Heterocyst, A Botanical Enigma." It opened with the following remark, "It is, to say the least, very unusual to find in a circumscribed group of plants a structure exhibiting a considerable degree of specialization for which, in the present state of our knowledge, it is impossible to suggest a really plausible function." The enigma no longer exists; the function of the heterocyst has been established. The challenge now is to try to understand the mechanistic details of how the various developmental events occur.

Chapter 6

Streptomyces

The authors of a recent review on the developmental biology of the actinomycetes opened their review with a startlingly frank comment: "This paper is not going to suggest a new model for studies in prokaryotic cell development. It is believed that the *Actinomycetales*, taken as a group, would present difficulties, given the current state of our knowledge, although they do pose intriguing and sometimes unique questions closely related to the area of developmental research."[1] The authors are quite correct in their implication that research on actinomycete development has lagged seriously behind that concerned with other prokaryotic systems. However, recent events suggest that this situation is rapidly changing. A number of major laboratories are now working with the actinomycetes, and it seems certain that rapid progress will be made. There are a number of reasons why the actinomycetes, or more accurately, the genus *Streptomyces*, is an appropriate subject for developmental analysis.

1. The morphological and developmental behavior of the group is complex and interesting.
2. Many of the actinomycetes, notably the genus *Streptomyces*, can be conveniently handled via conventional microbiological techniques.
3. There is available, for *Streptomyces coelicolor*, an excellent system for genetic analysis. This is due, largely, to the work of David Hopwood who has been a pioneer in the field. In addition, other laboratories have been working on the genetics of another species, *S. lividans*, and the results are quite promising.
4. The actinomycetes produce an elaborate variety of secondary metabolites. Thus, the applied and industrial consequences of understanding actinomycete development must be substantial.

Why then has understanding of this group lagged so? There are, I believe, two answers. Actinomycete development—namely, spore morphogenesis and the development and behavior of aerial mycelia—occurs mainly on the surface of a solid medium. Microbial physiologists are generally uncomfortable dealing with processes that do not take place in dispersed, liquid culture. Up until recently, most of the emphasis in prokaryotic development has been on endospores, myxospores, *Caulobacter*, heterocysts, azotocysts—all systems that can be studied in liquid culture. (Only recently has surface development of myxobacteria become amenable to convenient experimentation). Clearly, what have been needed to accelerate the developmental study of the actinomycetes are: (1) techniques for inducing the characteristic developmental changes in liquid suspension; and (2) techniques for doing biochemistry and molecular biology with surface cultures. These are challenges but neither is formidable. A variety of techniques that have been devised to study the behavior and development of *Dictyostelium* are available, and recent advances in studying aggregation and fruiting body formation in myxobacteria have been developed. Furthermore, it has recently been shown that *Streptomyces griseus* can be induced to form arthrospores in liquid medium.[2] While it is unfortunate that this is not the same species with which essentially all of the genetics has been done *(S. coelicolor)*, recent evidence suggests that it may be generalizable to other species of *Streptomyces*, and thus it may be feasible to study spore morphogenesis in *Streptomyces*.

The second reason is that much of the work that has been done on the physiology and biochemistry of *Streptomyces* has never found its way into the scientific literature. This is a result of the fact that most of the work has been done by commercial groups interested in antibiotic production and concerned primarily about proprietary interests.

A. General Description[3]

In a general sense, the actinomycetes are a group of rod-shaped, Gram-positive bacteria with a tendency toward a mycelial type of growth. Among the actinomycetes this mycelial tendency is exemplified in a subgroup referred to as the euactinomycetes. One member of this group is the genus *Actinoplanes*. This is a fascinating group of bacteria whose life style is strikingly similar to that of many fungi. In addition to their formation of a mycelial mat of cells, the ends of certain of the hyphae differentiate into sacs or sporangia containing a number of motile zoospores (Figures 6-1, 6-2, and 6-3). Upon release and eventual germination of the flagellated spores, a new vegetatively growing mycelial mat is generated. Unfortunately, these organisms are rather slow-growing and, at least at the present time, their behavior is unpredictable. Until there is enough additional information

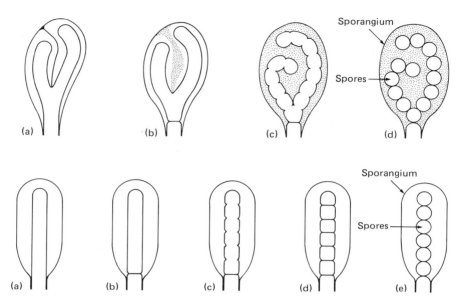

Figure 6-1 Diagram of spore and sporangial development in *Actinoplanes*. (Adapted from Lechevalier and Holbert 1965)

about their biology, this fascinating group of bacteria will continue to be ignored by most developmental microbiologists. Another member of the euactinomycetes, and one sufficiently domesticated to be a useful laboratory system, is the genus *Streptomyces*. *Streptomyces* has, in fact, been the subject of most of the work that has been done on the developmental biology of the actinomycetes and the rest of this chapter will focus on this genus.

Streptomyces is characterized by the formation of a branching, substrate mycelium that is occasionally interrupted by a cross-wall. When faced with nutrient depletion, the population begins to form aerial mycelia as branches of the substrate mycelia. At the ends of some of these aerial hyphae, the cells differentiate into a chain of spores. During the formation of aerial mycelia and spores, the population of substrate mycelia undergoes massive lysis. An interesting schematic representation of a colony of *Streptomyces* is presented in Figure 6-4. Any part of the population—spores or fragments of substrate or aerial hyphae—can give rise to a new colony.

B. Properties of the Spore

Figure 6-5 is a phase contrast photomicrograph of *S. coelicolor* illustrating the appearance of aerial mycelia and spores. A more detailed view of these

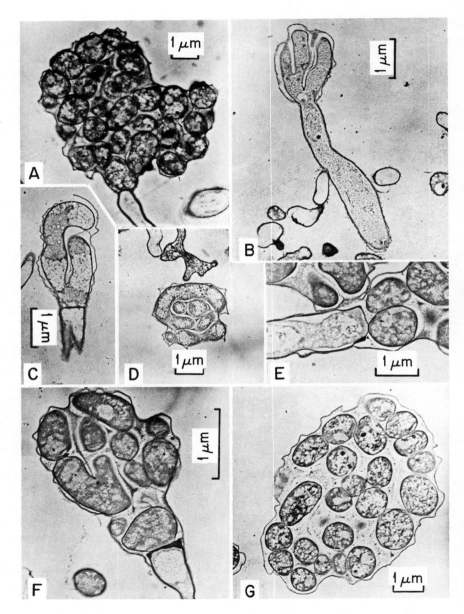

Figure 6-2 Electron micrographs of sections through a sporangium of *Actinoplanes* at various stages of development. *(A)* mature sporangium (4 days old); *(B)*, *(C)*, and *(D)* immature sporangium (2 days old); *(E)* and *(F)* mature sporangium (4 days old). Note that the sporangial wall is continuous with the sporangiophore sheath. *(G)* mature sporangium (4 days old). (From Lechevalier and Holbert 1965)

Figure 6-3 Electron micrograph of a thin section through an immature sporangium of *Actinoplanes*. (Courtesy of Dr. H. Lechevalier)

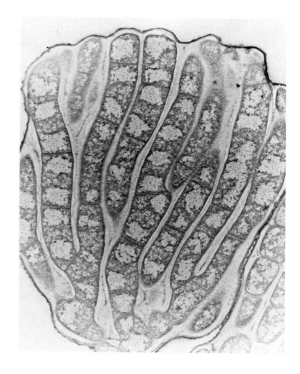

structures is presented in Figures 6-6 and 6-7 containing electron micrographs of thin sections of aerial mycelium and spores.

From a morphological point of view, it is not obvious that the hyphal spore is any more than a cell that has been pinched off from the aerial mycelium. While the wall of the spore is somewhat thicker than that of the mycelial cell, there seems to be no qualitative change in the wall peptidoglycan; nor are there any additional layers analogous to the cortex or spore coat of other types of resting cells. The spores are extremely hydrophobic; they resist being suspended in water and can be wetted only in the presence of detergent. They share this property with the aerial mycelium, and it seems to be attributable to a sheath that surrounds the cell wall. In many species of *Streptomyces*, the surface of the sheath is covered with spiny or hairy protruberances. Upon sonic oscillation, the sheath can be disassociated into submicroscopic rods or tubules. These seem to have the property of self-assembly and represent a common strategy used by organisms to construct external organelles that are not in direct contact with the cell membrane.

The spores are not especially heat-resistant, although they tolerate mildly elevated temperatures (e.g., 55 C) slightly more readily than do

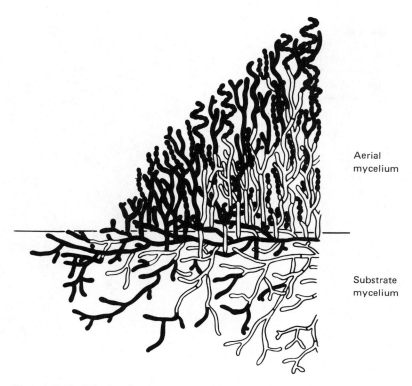

Figure 6-4 Idealized diagram of a vertical section through the center of a sporulating colony of *Streptomyces coelicolor*. Black represents intact cells and white represents disintegrating or lysing cells. (Adapted from Chater and Hopwood 1973)

the parent aerial mycelia. They are resistant to desiccation and have the important property of being metabolically dormant. Spores that are harvested in the dry state show no endogenous respiration, contain low levels of cytochromes and substantially less ATP than vegetative cells. Thus, the spores seem indeed to be resting cells.

The morphological events leading to the conversion of aerial mycelium to hyphal spores have been divided into four stages, illustrated diagrammatically in Figure 6-8. Stage 0 represents the vegetatively growing hyphae. At stage 1, presumably induced by depletion of some regulatory nutrient, the hyphae begin to assume the characteristic coiling. At stage 2, the hyphal strand is partitioned into individual cells by the appearance of septa or crosswalls. This stage was accompanied by DNA replication resulting in at least one genome per partitioned cell. Stage 3 involves the synthesis of new cell wall material, resulting in a thickened cell wall and the deposition of cell wall material on the developing septum. In stage 4, the spores round off and separate as the outer hyphal wall disintegrates.

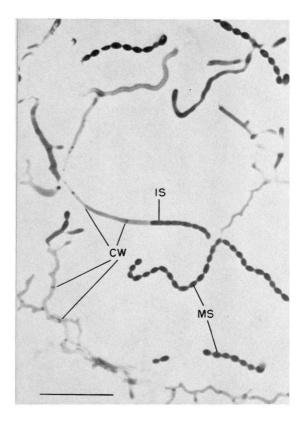

Figure 6-5 Phase contrast photomicrograph of *Streptomyces coelicolor*. Note the cross-walls *(cw)* in the vegetative hyphae, the rod-shaped immature spores *(is)*, and the helical chain of mature spores *(ms)*. The scale bar is 10 μm. (From Chater and Hopwood 1973)

The cells continue to be surrounded by the outer sheath. Some of these stages are illustrated in Figures 6-9, 6-10, and 6-11.

C. Spore Germination

The striking, but superficial, similarity between the actinomycetes and the fungi has led to a great deal of taxonomic and terminological confusion. As a result, following the fungal model, germination is often referred to as the process leading to the emergence of a germ tube from the spore. Even though the hyphal spores of *Streptomyces* are not closely similar to bacterial endospores, it is probably more useful and accurate to follow the terminology used for germination of *Bacillus* endospores. Thus, the overall germination process has been divided into three stages (see Chapter 3, Section D)—activation, initiation, and outgrowth.

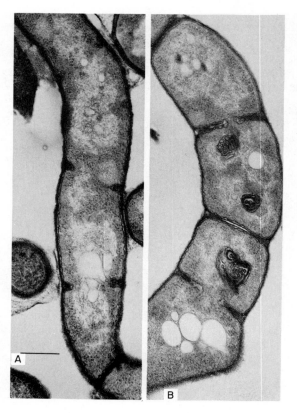

Figure 6-6 Electron micrographs of thin sections of *Streptomyces coelicolor* aerial mycelium during *(A)* the early stage of formation of sporulation septa and *(B)* the late stage of sporulation septation. The scale bar is 0.5 μm. (From Chater and Hopwood 1973)

As indicated in the discussion of endospore germination, activation is a necessary prelude for those spores that have not been allowed a period of aging. It is necessary in some *Streptomyces* and can be achieved by subjecting the spores to mild heating (e.g., 10 min at 55 C). Heating opens the door for the physiological changes that comprise initiation. As is the case for endospores, neither the mechanism nor the function of activation is understood.

Previously, studies of spore germination were carried out with spores that had been harvested under aqueous conditions from agar surfaces. The aqueous conditions (in combination with the nutrients leached out of the medium) must certainly have set some of the early germination events in motion, even if carried out at low temperatures. Recently, a technique has been developed for studying germination of *Streptomyces viridochromogenes* under much more controlled conditions.[4] The spores are harvested with dry glass beads and germination allowed to take place in a defined medium. Under these conditions, a requirement for activation was demonstrated, and initiation, which lasted 80 minutes, took place. During

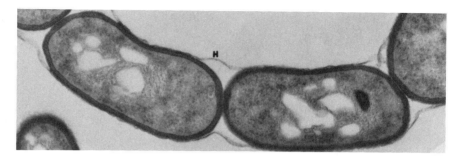

Figure 6-7 Electron micrograph of a thin section of mature spores of *Streptomyces coelicolor*. Note the somewhat thickened spore walls. (From McVittie 1974)

initiation, there was a 15% drop in optical density, reflecting the changed optical properties of the germinating spores, and all of the physiological events associated with germination took place. These included an increase in the rate of endogenous respiration, increased ATP content, and the initiation of RNA and protein biosynthesis. In other words, both the energetic and biosynthetic preparations for actual outgrowth were taking place. DNA synthesis then preceded the actual outgrowth during which the germ tubes emerging from the germinated spores developed into young vegetative hyphae. It is interesting to point out that the initiation phase of germination of *Streptomyces* is essentially a biosynthetic process, unlike the situation in *Bacillus* endospores where initiation is a degradative process. This is consistent with the fact that endospore germination is an endogenous process that can take place in the absence of external nutrients; *Streptomyces* spore germination, on the other hand, requires a variety of organic and inorganic nutrients.

The situation with regard to the mechanisms of dormancy and resistance in *Streptomyces* spores is even more bleak than in the case of *Bacillus* endospores; no insights at all are available.

D. Antibiotics[5]

Over two-thirds of all known antibiotics are produced by the streptomycetes. It seems obvious that there must be, therefore, some profound and fundamental relationship between antibiotics and the life style of the streptomycetes. The nature of this relationship is the basis of a debate that has gone on among microbiologists since the discovery of antibiotics. The two poles of argument are whether the primary function of an antibiotic is

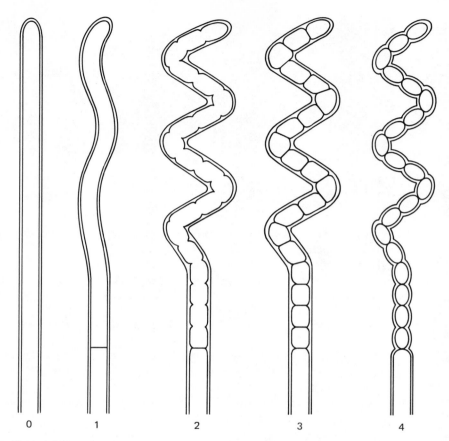

Figure 6-8 Diagram of the four stages of sporulation in *Streptomyces coelicolor*. After a phase of growth (0), the sporulating hyphae are divided into long cells by ordinary cross-walls, and the tips begin to coil (1). The apex is then partitioned into spore-sized compartments by sporulation septa (2). The cell walls thicken and constrictions appear between the young spores (3). As spores mature, they round off and separate (4). Some spores begin to germinate immediately after maturation. (Adapted from Wildermuth and Hopwood 1970)

an ecological one or a regulatory/developmental one. In the former sense, it has been argued that the excreted antibiotic generates a zone around the streptomycete that is inhospitable to other soil microorganisms, thus reducing competition for space and nutrients. An interesting variant of this idea is that the developmental lysis of substrate mycelia during development of aerial mycelia and spores generates a nutrient-rich local area that would be perceived and rapidly invaded by neighboring bacteria were it not for the inhibitory effect of the antibiotics.[5]

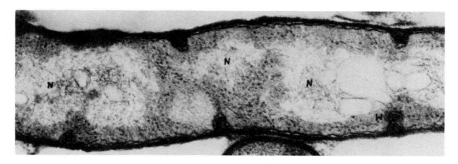

Figure 6-9 Electron micrograph of a thin section of sporulating hypha of *Streptomyces coelicolor*. This cell is at a very early stage of sporulation and shows two adjacent sporulation septa with ingrowths of the cell wall. N, nucleoplasm. (From McVittie 1974)

The alternative argument, the regulatory/developmental one, proposes that the antibiotics are regulatory molecules that are produced at specific stages of the developmental cycle; they may serve either to regulate and coordinate the regulatory processes involved in development or to maintain the metabolic dormancy of the spores. This argument is compromised by the fact that some antibiotic negative mutants of *S. coelicolor* are able to sporulate normally. It should be emphasized that a single strain of *Streptomyces* may produce multiple antibiotics and that one or more of these may serve multiple functions. Ensign has suggested that certain *Streptomyces* antibiotics may function as germination inhibitors. Germinating spores of *S. viridochromogenes* released an inhibitory factor similar in its properties to the antibiotic, nonactin. This antibiotic inhibited the germination of dormant or heat-activated spores of the same strain. Ensign proposes that this apparently self-defeating behavior of the organism actually serves to

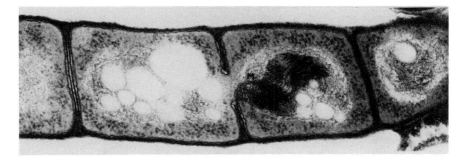

Figure 6-10 Electron micrograph of a thin section of a sporulating hypha of *Streptomyces coelicolor*. Notice the partially completed sporulation septum. (From McVittie 1974)

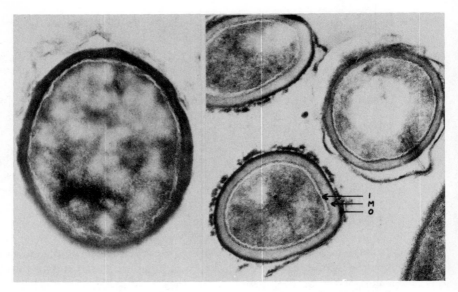

Figure 6-11 Electron micrograph of thin sections of mature spores of *Streptomyces coelicolor*; i, m, and o represent inner, middle, and outer layers of the spore wall. (From McVittie 1974)

prevent germination of other spores that might compete for limiting nutrient.[6]

The problem is difficult to resolve experimentally. On the one hand, it is difficult to subject the ecological argument to experimental test; on the other hand, the developmental argument based on the properties of antibioticless mutants must be viewed with caution. For example, what if this mutant produces essentially undetectable amounts of antibiotic sufficient, however, to exert a regulatory effect? Furthermore, the developmental effect of the antibiotic may be an extremely subtle one not manifested, for example, simply by the gross appearance or disappearance of spores. At this time, the question is simply not resolved.

E. Genetics[7]

Genetic analysis can play two roles: it can help to clarify the role of gene expression during development, and it can facilitate the analysis of the physiological processes during development. The various general strategies will be discussed more extensively in Chapter 12.

Up until recently, most of the genetic analysis of *Streptomyces* has been done in David Hopwood's laboratory and has focused on a single species, *S. coelicolor*. However, Stanley Cohen's lab at Stanford has been focusing its interest on developing systems for genetic analysis of another species, *S. lividans*. In order to do useful genetics, one wants to be able to do two things: to obtain mutants and to move DNA from one cell to another. The former is readily accomplished in *Streptomyces*; the latter is limited to the conjugal movement of plasmids between cells and to polyethylene glycol mediated protoplast fusion. This is an extremely efficient method for accomplishing recombination with efficiencies of 0.1 to 0.4 routinely obtainable; that is, 10 to 40% of the input cells undergo recombination. It is also possible to obtain genetic transformation with purified DNA using the polythylene glycol system. Efficiencies of 10^3 to 10^7 transformants/μg phage or plasmid DNA have been obtained. About 100 genes have been recognized and mapped; these include twelve or thirteen morphological/developmental genes. The genes are all on a single circular linkage map (Figure 6–12). Recently, there have been developed both a phage-mediated transduction system and a plasmid DNA-mediated transformation system. Thus, there are now available methods for a fine structure analysis of developmental genes as well as systems for cloning these genes, modifying them, moving them around, and generally making use of the powerful tools of genetic and molecular analysis.

In order to answer the question of whether a series of developmental genes is expressed in a linear, dependent fashion or separately on two or more connected pathways, a technique called *epistasis* was used. This method was originally developed for studying the sporulation genes in *Bacillus* and is based on the idea that if two genes are expressed on a linear, dependent pathway, then the phenotype of an appropriate double mutant will be that of one or the other gene, depending on which is expressed earlier. If, on the other hand, the two genes are expressed via different pathways, the phenotype of the double mutant should be unlike that of either mutant. Using this approach, it has been possible to analyze the orientation of a subset of developmental genes that control the "whi" phenotype. This stands for *white* and refers to the fact that as spores develop, the color of the aerial mycelium changes from white to gray. Mutations blocking spore formation prevent the colonial color change. Five of the eight whi loci were analyzed and were all found to lie on a linear dependent sequence, as illustrated in Figure 6–13.

Another set of developmental mutants are the "bald" mutants, referring to the fact that these mutants fail completely to synthesize aerial mycelia and thus lack the characteristic hairy or fuzzy colonial morphology (Figure 6–14). These have been grouped into three or four loci that have been shown not to be closely linked to each other. An examination of the

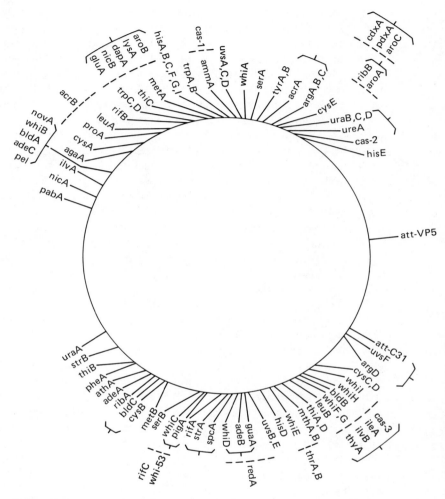

Figure 6-12 Genetic map of *Streptomyces coelicolor*. The order of those loci within brackets is undetermined; loci outside the dashed lines have not been ordered. (Adapted from Hopwood 1976)

linkage map, however, reveals that there do seem to be some clusters of developmental genes suggesting the possibility of some regulatory clustering (Figure 6-12).

It seems rather surprising that there are so few developmental genes in *Streptomyces*. You may recall that for *Bacillus* endospores it has been calculated that about 40 sporulation loci are involved. Since each locus may

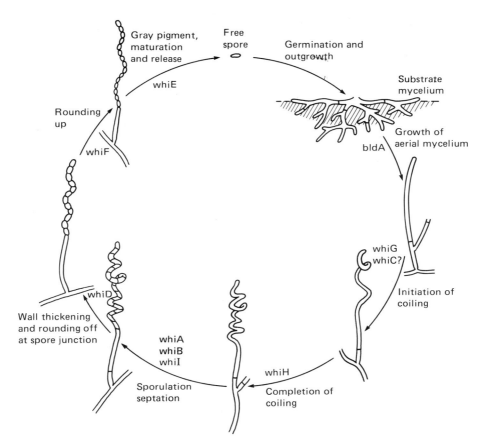

Figure 6-13 Genetic control of spore development in *Streptomyces coelicolor*. The symbols inside the arrows represent the genes thought to be involved at each particular developmental stage. (Adapted from Hopwood et al. 1973)

represent a contiguous cluster of genes, this figure is certainly a low estimate. Either development in *Streptomyces* is a far simpler process than in *Bacillus*, or there are, for some reason, a large number of as yet undiscovered genes. The latter seems more likely as the map is not yet saturated and the mapping procedures have relatively low resolution.

As is the case in *Bacillus*, *Streptomyces* genetics is hampered by the fact that the developmental mutants are only morphologically defined. Thus, there are no gene products associated with the various genes and, at the moment, this fact stands between the mutants and their analysis.

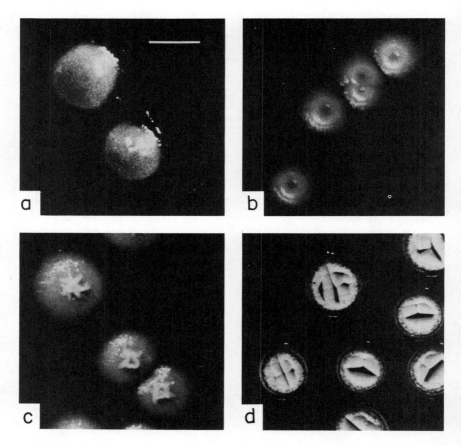

Figure 6-14 Colonial morphology of wild-type and bald mutants of *Streptomyces coelicolor:* *(a)* wild type; *(b)* class 1 bald mutant; *(c)* class 2 bald mutant; *(d)* class 3 bald mutant. The bar marker is 250 mm. (From Merrick 1976)

F. Regulation of Developmental Events

Among the actinomycetes, the absence of penetrating studies that proceed beyond the descriptive is nowhere more evident than when one tries to find molecular approaches to the regulation of development. Because of the difficulty in obtaining homogeneous or synchronously developing populations of aerial mycelia or spores, the approaches that have been reported are limited to the following:

1. *Kinetics of macromolecular synthesis during spore germination.*[8] A number of studies have been done with spores of different species of

Streptomyces during initiation and outgrowth. The conclusion on which there seems to be agreement are not unexpected; during the 60-minute period of initiation, RNA synthesis began immediately and continued at about the maximum rate throughout initiation. Protein synthesis began after about a 10-minute lag, reaching a maximum rate at the end of initiation. DNA synthesis began only during outgrowth (the emergence of the germ tubes from the spore). A number of experiments with rifampicin (which inhibits the initiation of RNA synthesis) showed that the antibiotic inhibited protein synthesis, implying that there was no stable mRNA involved in germination (i.e., RNA synthesis was necessary for continued protein synthesis).

2. *Comparative properties of ribosomes from spores and young vegetative mycelia.*[9] This approach is based on the notion that the ribosomes may play a role in developmental regulation. In other words, it examines the possibility that there is a translational level of control of protein synthesis. Let us say, for example, that the spore contains a stored, stable messenger RNA that plays a role in the initial steps of germination and therefore must not be translated while the spore is dormant. One might expect that this delayed translation could be accomplished by some modification of the ribosomal structure that prevents it from entering into a polysomal complex with the stable mRNA. In fact, in two cases when this has been examined, the spore ribosomes did indeed differ from those in the vegetative mycelia. Polysomes were absent, the spore ribosomes were more stable at low Mg^{++} concentrations (which usually dissociates the 70S ribosome to the 50S and 30S subunits), and in both cases the spore ribosomes seemed to contain a tightly bound pigment. While these sorts of results are interesting, they are also open to alternative explanations. For example, the 70S ribosome goes through a normal dissociation-association cycle during protein synthesis; when the 70S particle is released from the polysome during active translation, it dissociates into its 50S and 30S subunits, which must reassociate in order to form the initiation complex with the nRNA. If there is no mRNA available in the spore, there would, of course, be no polysomes, no protein synthesis, and the association-dissociation property of the ribosome (Mg^{++} dependent?) could be modified so as to stabilize the 70S particle.

3. *The developmental properties of rifampicin-resistant mutants.* This approach was first used to demonstrate that spore development in *Bacillus* involved a control mechanism at the level of the RNA polymerase.[10] Rifampicin is an antibiotic that inhibits RNA synthesis (of all types—mRNA, tRNA, and rRNA) by binding to the

β-subunit of the RNA polymerase and preventing initiation of transcription. The rationale of the experiment was as follows: if the RNA polymerase of vegetative cells is only able to recognize the promoters of genes involved in vegetative growth, and if sporulation involves a modification of the polymerase structure or subunit composition that now allows the polymerase to recognize sporulation promoters, since rifampicin resistance is achieved by a structural modification of the polymerase that prevents it from binding the antibiotic, that same modification may prevent the polymerase from recognizing the sporulation promoters. Thus, the prediction is that some fraction of rifampicin-resistant mutants should also be developmentally deficient. This was indeed the case in *Bacillus* (see Chapter 3, Section H) but unfortunately not in *Streptomyces*. However, the experiment illustrates an interesting pitfall of this approach. One hundred rifampicin-resistant mutants were obtained by mutagenesis; some were unable to form aerial mycelia. If that had been the end of the experiment, one would have concluded that there was indeed a developmental consequence of polymerase modification. However, the entire thesis rested on the assumption that both the rifampicin resistance and the developmental block were expressions of a change in a single gene. Since any mutation induced by mutagenesis and, in fact, even spontaneous mutations are likely to be accompanied by other mutations, it is thus necessary to demonstrate that the two effects are actually attributable to a single gene. This can be accomplished either by obtaining a back mutant and determining whether both properties are changed (the probability that this will occur simultaneously in two genes is vanishingly low) or by back-crossing the mutated gene into a wild-type background and determining if the two properties are transferred together. When crosses were done with the *Streptomyces* mutants, the two properties segregated, indicating that the polymerase modification had no developmental consequences. The conclusions that may be drawn from these results are either that changes in RNA polymerase are not involved in developmental regulation in *S. coelecolor* or that such changes, if they do occur, are at a portion of the molecule distinct from the rifampicin binding site.[11]

G. Streptomyces DNA

The DNA of the *Streptomyces* is rather unusual. First, it is extremely high in %G + C content, ranging from 63 to 72%. It shares this property with a few other bacteria, for example, the entire group of the myxobacteria. Second, the genome size is extremely large for a prokaryote. The values of

7.1×10^9 daltons for *S. coelicolor* and 6.8×10^9 daltons for *S. rinosus* are three times the size of the genome of *Escherichia coli* and are among the highest in the prokaryotes.[12] Only the cyanobacteria with genome sizes up to 8×10^9 daltons share this property. It is difficult to imagine what the function of so much extra DNA is. The genome of *Bacillus* is only about 2.0×10^9 daltons even though it must code for the endospore. Third, the complexity of the *Streptomyces* genome is interesting. In *S. coelicolor*, 2% of the genome is "foldback" DNA, and about 5% consists of repetitive sequences with about four copies per haploid genome. The foldback DNA reflects the presence of inverted repeat sequences and recalls the fact that in *Caulobacter*, also, 3% of the DNA was present as inverted repeats. Whether or not these sequences in *Streptomyces* are moveable or invertible has not been examined but, as in the case with flagellar phase variation in *Salmonella*, offers the possibility of the sort of metastable regulatory device that would be appropriate for developmental controls.

H. Extracellular Factors

One characteristic of some multicellular eukaryotes is the ability of cells to communicate by means of the exchange of extracellular signals. In prokaryotes, this phenomenon is limited to a few organisms and will be discussed more fully in Chapter 11. The two developing prokaryotes that manifest signal exchange are the myxobacteria (see Chapter 7) and *Streptomyces*. In both systems, the experimental strategy for examining these signals has been to try to correct a developmental lesion in a nondeveloping mutant by the addition of material from a parent strain. The developmental rescue is symptomatic and nongenetic; in the absence of the added signal, the mutant retains its genetic inability to complete development. It is interesting and certainly relevant that both myxobacteria and *Streptomyces* also manifest what could be thought of as a primitively multicellular life style.

Bald mutants of *S. griseus* (mutants that produce no aerial mycelia or spores) regain their full developmental capacity when exposed to small amounts of the culture filtrate of the parent developmentally competent strain. The active material has been called "factor A," and its chemical structure has been determined to be 2[S]-isocapryloyl-3[R]-hydroxymethyl-butyrolactone ($C_{13}H_{22}O_4$). This is a biologically strange molecule but there are some data that suggest a function and mechanism. Apparently, the A factor stimulates a NAD glycohydrolase and, as a consequence of the lowered levels of cellular NAD, a higher proportion of substrate glucose is shifted to the biosynthesis of streptomycin and correspondingly less to growth and macromolecular biosynthesis. A gene for factor A has recently been cloned and, upon introduction into A factor-minus mutants of *S.*

griseus, restores to the mutant the ability to produce streptomycin, resistance to streptomycin and spore formation. It appears that the gene(s) for factor A are carried on an unstable, extrachromosomal element.[13]

Another cellular signal also involved in the development of *S. griseus* is referred to as "C factor."[14] Again, material produced during the growth of a developmentally competent strain can phenotypically complement a mutant unable to form spores. C factor has been partially purified and seems to be a 34.5 K dalton peptide. There are conflicting data concerning its mechanism of action, but it has been claimed to play a role in regulating RNS synthesis.

I. Concluding Remarks and Salient Questions

With *Streptomyces* one hardly knows where to begin. It seems obvious that one of the most serious impediments to the investigation of *Streptomyces* development has been the failure to devise experimental systems that clearly and conveniently separate the various developmental events one from the other. The formation of vegetative and aerial mycelia must be separated, as must the formation of hyphal spores from the aerial hyphae. (It is already possible to study spore germination with clean preparations of dormant spores, and there are unpublished reports of bona fide spore formation in liquid suspension). Such separations would then allow the processes to be characterized physiologically. What structural macromolecules are changed? What are the metabolic, enzymatic changes? The various developmental mutants must be characterized and distinguished in terms of gene products rather than broad morphological criteria. The existence of the lysogenic and transducing phage, ΦC31, has partially set the stage for a molecular and genetic analysis of the developmental events. In addition, the actinomycetes, like the myxobacteria, offer the opportunity to examine primitive cell interactions in a prokaryotic system.

At the present time, most of the genetic studies with *Streptomyces* are focused on antibiotic formation. It is tacitly assumed that the antibiotics are developmentally relevant and, in fact, play a functional role in the *Streptomyces* life cycle. Thus, it is implied that the study of the genetics of antibiotic formation is congruent with the study of the genetics of development. This may or may not be so. In any case, there has been a long-standing interest in the role of secondary metabolites in plants and fungi. It is not overly optimistic to hope that this issue may yet be settled through the study of *Streptomyces*.

Finally, it is encouraging that at least two major laboratories have recently redirected a portion of their activities to examining molecular aspects of regulation in *Streptomyces*.[15] Their efforts will certainly help to redress the neglect that these organisms have been subject to in the past.

Chapter 7

The Myxobacteria

"A few years since, while collecting fungi at Kittery and in several localities in New England and the Southern states, the writer's attention was attracted by a bright orange colored growth occurring upon decaying wood, fungi and similar substance."[1] Thus, Roland Thaxter in 1892 first described what he called the "Myxobacteriaceae" that were subsequently to be referred to colloquially as the myxobacteria (Figure 7-1 and 7-2). Twelve years later, writing in the same journal, Thaxter grumbled, "Since the appearance in this journal (June 1897) of the author's second paper on this group, comparatively little attention has been given to it, and with few exceptions one finds no mention of it in current text-books or in bacteriological literature."[2] Today, Professor Thaxter would be pleased to discover descriptions of the myxobacteria in almost any microbiology textbook and over 800 papers in the literature on various aspects of their biology and biochemistry.

In 1924, the German botanist E. Jahn[3] published his now classic monograph on the myxobacteria *(Die Polyangiden)*, and from then on the myxobacteria have been firmly in place among the bacteria. However, they remained burdened by the stereotype of being exotic and difficult to work with. Their reputation reflected, in part, the fact that in liquid culture they grew in clumps as pellets and attached to the walls of the vessel, rather than in the dispersed stage bacteriologists are so fond of. In 1948 Nellie Angelina Woods, who has remained an unsung heroine of myxobacterial research, described a dispersed-growing strain of *Myxococcus xanthus* in her master's thesis.[4] Since then, many such strains have been isolated from different genera, and one can date the resurgence of myxobacterial research from that point on. By now, many basic aspects of the biology of the myxobacteria have been described, and it is fair to say that the stage is now set for a detailed analysis of their developmental behavior.[5] Attention, however, has

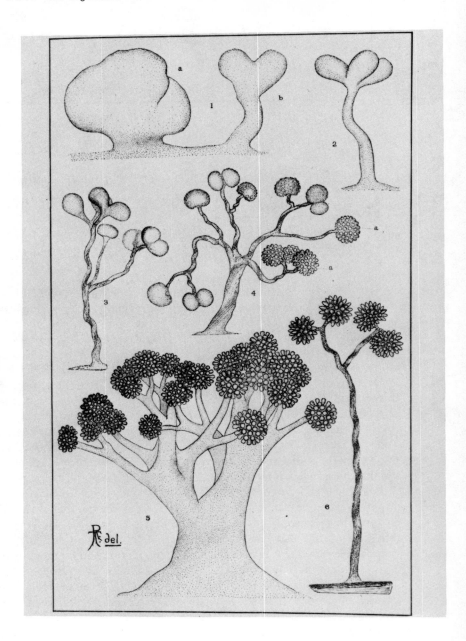

Figure 7-1 Roland Thaxter's original illustration of the fruiting bodies and the cell morphologies of a variety of myxobacteria he had isolated. (From Thaxter 1892)

Chapter 7 The Myxobacteria 107

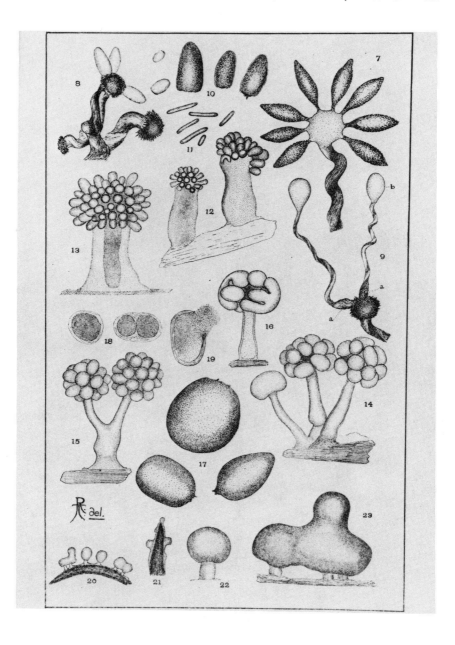

Figure 7-2 Roland Thaxter's original illustration of the fruiting bodies and the cell morphologies of a variety of myxobacteria he had isolated. (From Thaxter 1897)

Chapter 7 The Myxobacteria

focused primarily on one organism, *M. xanthus*, and recently on a second, *Stigmatella aurantiaca*. This chapter will deal primarily with those organisms.

A. The Life Cycle

The myxobacteria are Gram-negative, rod-shaped soil bacteria that are distinguished from other bacteria by two properties, one unusual and the other unique. The unusual property is their ability to move by gliding over a solid surface. They share this property with a number of other bacteria, colloquially referred to as the gliding bacteria. The feature of the myxobacteria that distinguishes them from all other bacteria is their life cycle. In reality, it consists of two alternative cycles—one of growth and the other of development. In the growth cycle, Gram-negative rods grow and divide by conventional binary, transverse fission until growth becomes limited by nutrient exhaustion or by end product accumulation (growth and nutrition will be discussed in a later section). If three conditions are met, the cells shift to the alternative cycle of development:

1. The cells must sense an appropriate nutrient depletion.
2. The cells must be on a solid surface.
3. The cells must be at a sufficiently high cell density.

These three requirements will be discussed in greater detail in a later chapter.

The life cycles of *M. xanthus* and *S. aurantiaca* are diagrammatically illustrated in Figures 7-3 and 7-4. Approximately when growth has ceased, the cells aggregate by an unknown mechanism into aggregation centers that serve as focal points for the development of the fruiting bodies. During aggregation, there is massive lysis of 40 to 90% of the cells, the number depending on the conditions and the organism. The aggregates are eventually converted to fruiting bodies that are simple, elevated mounds of about 10^4 cells (Figure 7-5) or elaborate, differentiated structures such as *Chondromyces* (Figure 7-6) or *Stigmatella* (Figure 7-7). The vegetative cells undergo a cellular morphogenesis to resting cells that are somewhat resistant, metabolically quiescent cells called *myxospores* (Figure 7-8). In the case of some genera, such as *Myxococcus*, these are morphologically altered to round, optically refractile cells with a resistant spore coat. In other genera, whose physiological properties have hardly been studied, the resting cells undergo little morphological conversion. In genera, such as *Myxococcus* and *Chondrococcus*, the spores are not contained in any kind of sac and are themselves the unit resting cells; while in other genera such as

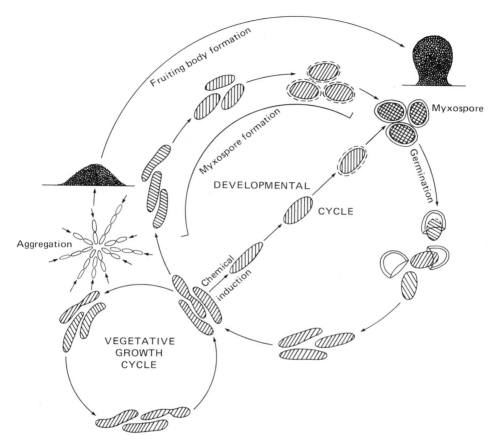

Figure 7-3 Diagram of the life cycle of *Myxococcus xanthus*. The fruiting body is not drawn to scale but is a few hundredths of a mm in diameter, in contrast to the vegetative cells, which are about 5 to 7 × 0.7 μm.

Stigmatella or *Chondromyces*, the spores are contained in large cysts called *sporangioles* (Figure 7-9). In these myxobacteria, the sporangiole is probably the unit resting structure. Myxospore germination is induced by favorable nutritional and physical conditions, and in organisms such as *Myxococcus* individual myxospores germinate and are converted back to vegetative rods. In organisms such as *Stigmatella*, where the spores are contained in a sporangiole, germination involves the bursting of sporangiole and the morphogenesis of the slightly shortened rods back to the vegetative cells.

In *Myxococcus*, the entire process can be short-circuited and the cells induced to convert directly to myxospores by the addition to an actively growing culture of high concentrations of any one of a number of chemical

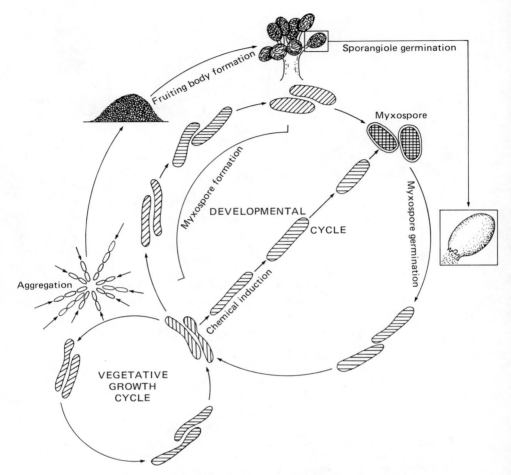

Figure 7-4 Diagram of the life cycle of *Stigmatella aurantiaca*.

inducers (e.g. 0.5 M glycerol). The induction is rapid (ca 90 min), quite synchronous, and relatively complete (i.e., usually 95% of the cells are converted). Glycerol induction will be discussed later in more detail.

B. Ecology

The myxobacteria are soil bacteria, although they have been isolated from a variety of other sources where they may have been only transient residents. There have been a number of attempts to characterize the distribution of the

Figure 7-5 Photograph of the fruiting body of *Myxococcus fulvus*. (From Reichenbach and Dworkin 1981)

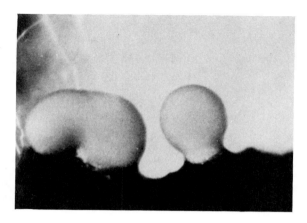

myxobacteria; all of these have failed, however, because of the difficulty of determining myxobacterial numbers by conventional techniques. Myxobacteria exist in nature either as a swarm of vegetatively growing cells surrounded by polysaccharide slime or as myxospores embedded in a fruiting body. In either case there are two problems, one technical and the other philosophical. The technical problem is that it is extremely difficult to disperse the cells in the swarm or the myxospores in the fruiting body completely in order to carry out the cell count. The other problem is deciding in what terms to view the distribution of these organisms, given their quasi-multicellular state. If one finds, for example, one swarm or one fruiting body, each containing 10^4 cells, in a gram of soil, are there 10^4 myxobacteria/gm of soil or one? Obviously, there are both, and techniques are needed to allow both single-cell and swarm or fruiting-body determinations to be made.

There is little hard and useful information on the ecology of the myxobacteria.[6] However, some conclusions can be deduced or speculations made based on what has been learned in the laboratory. One of the very characteristic features of the myxobacteria is their ability to excrete hydrolytic enzymes and, as a result, to degrade and feed on extracellular molecules. Thus, all of the myxobacteria are able to lyse and feed on bacteria, living or dead. A powerful battery of peptidoglycan-lytic enzymes are excreted and, in the case of Gram-positive bacteria, can lyse living or dead cells at a distance from the myxobacterial cell. In the case of Gram-negative bacteria, the outer membrane prevents the lytic enzymes of the myxobacteria from gaining access to the cells' peptidoglycan. Such cells can be lysed only if the outer membrane is disrupted by heat or by an appropriate solvent. If, however, the myxobacterial cell comes in direct physical contact with the prey, it is able to lyse the cell. Apparently the myxobacteria contain some cell-surface bound enzymes that can lyse the outer

Figure 7-6 Fruiting bodies of *Chondromyces*. (A) *C. crocatus*, photographed with oblique illumination. The scale bar is 125 μm. (B) *C. apiculatus*. The scale bar is 150μm. Photographed through phase contrast. (Courtesy of Dr. Hans Reichenbach)

membrane of the Gram-negative cell. Myxobacteria also excrete proteolytic and saccharolytic enzymes; and a few myxobacteria are cellulolytic.

Thus, it seems fair to conclude that the myxobacteria serve in nature as scavengers of insoluble, macromolecular debris. If this is so, it allows one to make sense of the gliding motility and the life cycle. On the face of it, if one knew that a bacterium depended for its food on the solubilization of particulate, macromolecular debris, one could make certain sensible predictions about that organism's behavior. One might predict, for example, that its motility would not be a swimming type, ideally suited for maximizing contact with dissolved molecules, but rather that it might be of a gliding type suitable for crawling over and adhering to particulate material. Furthermore, one might guess that an organism whose nutrients are generated by the action of an extracellular enzyme has some peculiar problems. Its excreted enzymes are diffusing away from the cell, their concentration decreasing with great rapidity and as the square of the distance from the source. When the enzyme reaches its substrate, the nutrients generated by its action also diffuse away from the food source, their concentration also decreasing as the square of the distance. Thus, the bacterium is at the mercy of the forces of diffusion and dispersion. The strategy adopted by the

Figure 7-7 Fruiting body of *Stigmatella aurantiaca*. The scale bar is 40 μm. Photographed with oblique illumination. (Courtesy of Dr. Hans Reichenbach)

Figure 7-8 Myxobacterial myxospores. *(A)* Phase contrast photomicrography of the free spores of *Myxococcus xanthus*. *(B)* Phase contrast photomicrography of induced myxospores of *Stigmatella aurantiaca*. (Courtesy of Dr. Hans Reichenbach)

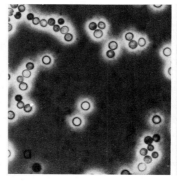 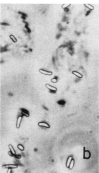

myxobacteria is to concentrate the cells into a swarm, thereby maximizing the efficiency of the excreted hydrolytic enzymes—in other words, a kind of microbial wolf-pack effect. That this, indeed, is the case has been demonstrated experimentally and mathematically,[7] and it seems to provide a rationale for the myxobacterial life cycle. The swarm optimizes feeding; thus, everything about the organism is designed to maintain a high cell

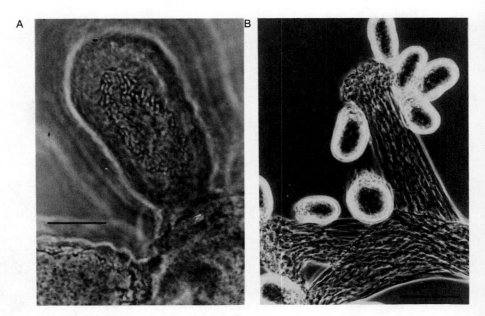

Figure 7-9 Sporangioles of myxobacteria. *(A) Stigmatella aurantiaca.* The myxospores are visible under the sporangiole. The scale bar is 12 μm.
(B) Chondromyces crocatus. Some sporangioles are attached and others have broken off the stalk. The scale bar is 32 μm. Both of these pictures were taken with phase contrast microscopy. (Courtesy of Dr. Hans Reichenbach)

density—the undefined attractive forces that keep the vegetative cells together in a swarm and the aggregation and fruiting body formation that maintains the resting cells at a high cell density so that, upon germination, a swarm is immediately present.

A variety of fascinating ecological questions beg to be answered: What are the differences in the distribution of myxobacteria with different developmental strategies, for example, free spores versus sporangioles (Figure 7-8 and 7-9), or fruiting bodies with pedestals versus those that rest directly on the surface (Figures 7-5 and 7-10)? What is the ecological significance of these differences? What is the role of the myxobacterial pigments? Under what conditions do myxobacteria exist as vegetative cells or resting cells? What are the effects of light-dark cycles, ambient temperature, humidity, food availability on the developmental cycle *au naturel?* What role, qualitatively and quantitatively, do the myxobacteria play in organic transformations? Finally, does the ability of the myxobacteria to kill and lyse other bacteria play any role in the competitive interactions that occur in the soil microflora?

It is most interesting that, in general, there is little attention paid to the ecological aspects of bacterial life cycles. The myxobacteria offer the

Figure 7-10 Fruiting body of *Myxococcus stipitatus*. (Reichenbach and Dworkin 1981)

extremely attractive and realistic possibility of examining the interaction between the natural environment and an organism's developmental cycle.

C. Taxonomy

The taxonomic state of the myxobacteria is unsettled and has been so since their discovery. Roland Thaxter was prescient when he commented, "The literary history of the *Myxobacteriaceae* [sic] thus bids fair to become as remarkable in its diversity as are the characters which make the order an anomaly among the plants which seem to be its nearest allies . . . and one can but look with no small interest, and perhaps with some misgivings to such further taxonomic vicissitudes as may be in store for them."[8]

The features that unite the myxobacteria are:

1. They are all Gram-negative rods even though there seem to be two lines of descent within the group—one consisting of organisms with long, thin rod-shaped cells and the other of short, blunt-ended rods.
2. The are all gliding bacteria.
3. They all feed by collecting the products of excreted hydrolytic enzymes.
4. They all form high density aggregates of cells which then differentiate into fruiting bodies of varying shape and complexity.
5. Most of them form resting cells; these vary from resistant, round spores to somewhat shortened rods.
6. The %G + C values for all of them fall between 68 and 72.

The principal reason for the taxonomic uncertainty is that only a handful of species has been examined from a physiological, biochemical, and molecular point of view sufficient to generate the objective parameters necessary for good taxonomy. The morphological and developmental complexity of the myxobacteria has been sufficient to allow for a first approximation of a taxonomic scheme. However, the subjective quality of such parameters makes the choice between alternative schemes a matter of taste rather than analysis. Nevertheless, it is generally agreed that the myxobacteria fall into a single order, the Myxobacterales, and that the order is further subdivided into four families. In one taxonomic scheme, the two lines of vegetative cell morphology are acknowledged by a division of the Myxobacterales into two suborders, the Cystobacterineae and the Sorangineae. The former contain the long thin rods and is divided into the families Myxococcaceae, Archangiaciae, and Cystobacteraceae. The latter contains the blunt, shorter rods and includes a single family, the Sorangiaceae.

D. Cultivation

Whereas freshly isolated strains of myxobacteria do not usually grow in a dispersed state in liquid media, strains that will do so can easily be selected; and it is those strains of *M. xanthus* and *S. aurantiaca* that are commonly used for laboratory studies. In *M. xanthus*, the basis for whether or not a strain grows in a dispersed state is correlated with the nature of its motility and will be discussed later in more detail; at this point, suffice it to say that these dispersed-growing strains are developmentally competent and thus are ideal for experimental manipulation.

With the exception of those few myxobacteria that are able to degrade cellulose, the myxobacteria require organic nitrogen sources. These requirements can be satisfied either by protein (e.g., casein), peptides (e.g., Casitone or Tryptone) or appropriate mixtures of amino acids. *M. xanthus* growing on a Casitone-salts medium has no unusual characteristics and grows with a generation time of 3.5 to 4 hours. It can also be grown on defined media containing amino acids and salts and under these conditions has a generation time of 8 to 12 hours. Most myxobacteria, including *M. xanthus* and *S. aurantiaca*, can also be grown on or in solid media, on a complex medium, and will form colonies from single cells. Thus, the usual sorts of laboratory manipulations such as obtaining isolated colonies for cloning, counting cells, and mutant selection, growing lawns for phage growth, and so on, can be routinely done. Whether the organisms grow vegetatively or, alternatively, go through the developmental cycle is controlled by the nature of the nutritional environment; at high nutrient concentrations development is inhibited and the cells simply grow and divide. If the level of nutrient is lowered appropriately, aggregation, fruiting-body formation, and sporulation take place.

It is useful to emphasize that growth of M. xanthus or S. aurantiaca poses no special problems. While the generation time does not match that of such sprinters as Escherichia coli or Pseudomonas, overnight cultures are quite practicable and substantial quantities of cell material are easily obtainable.

The myxobacteria have long suffered from the unjust stereotype that they are exotic, temperamental, or generally difficult to deal with. The ease with which even biochemists and molecular biologists have been able to work with myxobacteria should finally put this misapprehension to rest.

E. Cell Structure

1. Vegetative Cells

Both of the myxobacteria with which this chapter is concerned fall within one of the two lines of cellular morphology referred to earlier (the Cystobacterineae). While there is some variation within the group, the vegetative rods are about 5 × 0.7 µm (Figure 7-11). The fine structure of the cells is not unusual; they look like typical Gram-negative bacteria (Figure 7-12) with an outer membrane and a cytoplasmic membrane surrounding a thin layer of peptidoglycan. The surface layers of M. xanthus have been characterized; the outer membrane is more or less typical, containing protein, phospholipid, and lipopolysaccharide. The phospholipid is mainly phosphatidylethanolamine and is present in usually large amounts.[9] The nature of the outer membrane, and in particular its surface macromolecules, are of especial interest, since it is likely that the outer membrane will contain the various receptor sites for both physical cell-surface interactions and the extracellular signals exchanged by the cells during development.

The cells contain somewhat less peptidoglycan than does E. coli, and in addition, the cellular organization differs in a rather interesting way. The peptidoglycan seems to exist in patches that are held together by some trypsin-sensitive material.[10] Whether this has to do with the somewhat greater flexibility of the cells, with their gliding motility, or with their rod-sphere conversion during myxospore morphogenesis is unknown.

M. xanthus has been shown to be piliated (Figure 7-13);[11] furthermore, the pili have been shown to play an interesting role in one aspect of the cell's motility. This role will be discussed in more detail in Section G of this chapter.

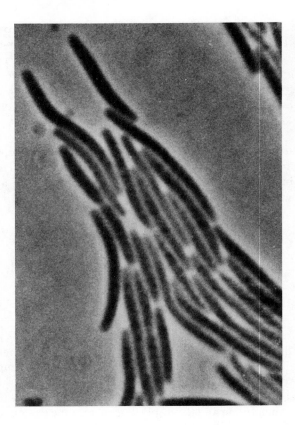

Figure 7-11 Phase contrast micrograph of vegetative cells of *Myxococcus xanthus*. Magnification is 2600×. (Courtesy of Dr. Hans Reichenbach)

2. Myxospores

The myxospore of *M. xanthus* and *S. aurantiaca* are illustrated in Figure 7-8. The myxospore of *S. aurantiaca* is contained within a sporangiole and is a refractile, somewhat shortened version of the vegetative cell. Neither its chemical makeup nor its physiological properties (resistance, metabolic rate, etc.) have been investigated. The myxospore of *M. xanthus* has been shown to be a metabolically dormant, resistant cell that is formed late in the process of fruiting-body morphogenesis.

The chemical composition of the myxospore peptidoglycan does not differ from that of the vegetative cell; however, there do seem to be organizational differences. There is a substantial increase in the myxospore, in the degree of cross-linking within the peptidoglycan; and unlike the vegetative peptidoglycan, it is no longer solubilized by trypsin and detergent. Whether this reflects a change in the peptidoglycan-trypsin sensitive complex itself or in its interaction with other cellular components of the spore surface is not known.[12] Most of these physiological studies of myxospores have been done

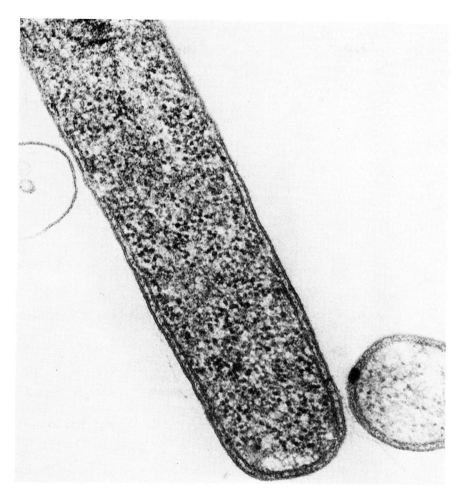

Figure 7-12 Electron micrograph of a thin section of a vegetative cell of *Myxococcus xanthus*. (Courtesy of Dr. Herbert Voelz)

with spores generated by the glycerol induction technique.[13] This technique will be discussed in greater detail later (see section J of this chapter); at this point, it is only necessary to point out that it is a technique for inducing sporulation by short-circuiting the normal developmental process; instead of being formed within fruiting bodies, the spores are induced in liquid culture, in a rather quick and convenient fashion, by the addition of one of a variety of inducers. While the spores thus formed are quite similar to fruiting-body spores, there are some important differences; and any time glycerol-induced spores are used as a model system, it is necessary that the

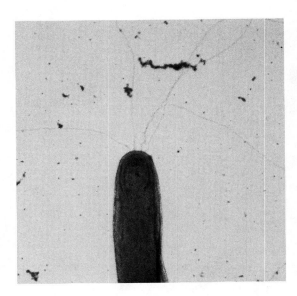

Figure 7-13 Electron micrograph of a stained cell of *Myxococcus xanthus* showing the polar piliation. (From Kaiser 1979)

property or process being examined be shown to be similar to that occurring normally.

The nature of the spore coat has been examined; its morphological fine structure is illustrated in Figure 7-14. In both fruiting-body spores and glycerol-induced spores, it consists largely of polysaccharide (75% by weight) with substantial amounts of protein (14%) and glycine (7%). Based on the inability to separate the protein and polysaccharide components of the coat without resorting to drastic hydrolysis procedures, it has been suggested that they are covalently bound to each other, perhaps as glycoprotein.[14]

A most interesting aspect of myxospore structure is the presence in mature fruiting-body myxospores (but not in glycerol-induced spores) of a protein called "protein S." It appears in an outermost 30 nm layer formed late in the fruiting process and has the additional interesting feature that it may be stripped off the myxospore and, upon readdition to the uncoated spore, self-assembles into the 30 nm layer.

Recent work from Masayori Inouye's lab has indicated that a portion of the gene that codes for protein S has a striking homology with the portion of the gene for bovine brain calmodulin that codes for the Ca^{++} binding segment of calmodulin. While this is consistent with the observation that protein S interacts with Ca^{++}, the function of protein S is unknown. The DNA that codes for protein S has been cloned; this revealed the unexpected finding that there were two closely linked, almost completely homologous genes (gene 1 and gene 2) that coded for protein S. Using the transposon Tn5, site specific mutations have been inserted in these genes in order to examine

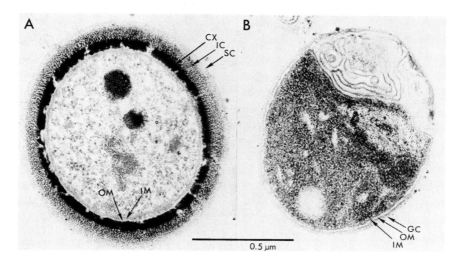

Figure 7-14 Electron micrographs of thin sections of myxospores from *Myxococcus xanthus*. (A) myxospore obtained from a fruiting body. (B) myxospore obtained by glycerol induction. OM, outer membrane; IM, inner membrane; CX, cortex; IC, intermediate coat; SC, surface coat; GC, glycerol spore coat. (From Inouye et al. 1979)

their function. In addition, they have been fused with promoterless *lacZ* genes to examine gene expression. It was found that gene 2 is responsible for almost all of protein S synthesis during development and that strains carrying insertions in either gene 1 or gene 2 aggregated and sporulated normally.[15] This is an exciting and puzzling aspect of myxobacterial development and illustrates the experimental power of the new approaches to genetics.

F. Myxobacterial Polysaccharide

The word *myxobacteria* is derived from the Greek *myxa* referring to a kind of slime. This derivation reflects the fact that myxobacteria characteristically produce extracellular polysaccharide during growth and development. The functions of the slime have not been investigated. However, it is produced in significant amounts at so many different points during the life of the organism that it is difficult to escape the conclusion that it plays a number of key roles.

Slime is produced and excreted by cells during vegetative growth, either in liquid culture or on the surface of agar plates. Under these conditions, the slime, in one case, amounted to somewhat over 10% of the dry weight of the cells. It can also be isolated from fruiting bodies of M. xanthus (and other myxobacteria) where it exists as an extracellular matrix loosely associated with the cells. Here the amount of slime produced was almost 20% of the dry weight of the cells.

While the nature of the myxobacterial polysaccharides has not been determined, their elemental composition in M. xanthus has been analyzed.[15A] The constituents of the vegetative cell polysaccharide(s) are mannose, glucose, and glucosamine. The fruiting-body polysaccharide(s) is composed of the same sugars with additional traces of galactosamine. (Both of the amino sugars may be acetylated; that has not been determined.) The ratios of the various sugars are somewhat similar for both vegetative cells and fruiting bodies. However, in the latter case the amount of glucose was considerably higher, perhaps owing to the fact that myxospores contain a considerable amount of an α-1,3 glucan as a surface layer.[16] Likewise, the myxospores contain substantial amounts of galactosamine as a component of the spore coat.[17] Whether this is responsible for the small amounts of galactosamine in the fruiting-body slime cannot be decided at this time. If these two explanations of the difference between the vegetative cell and fruiting-body slime are correct, one is left with the surprising and encouraging tentative conclusion that the growing and developing cells both produce the same polysaccharide(s). It is encouraging from a technical point of view in that one may collect and purify the polysaccharide much more conveniently from cells growing dispersed in liquid culture than from fruiting bodies.

I have spent what the reader might consider to be more space than is necessary on a rather unglamorous aspect of myxobacterial biology. However, I think that the myxobacterial polysaccharide is central to so many aspects of the biology of the organism that I am puzzled and distressed that so little has been done to characterize it. The following are some of the ways in which it may play a role in myxobacterial development:

1. It has been suggested that the slime is necessary for, or facilitates, gliding motility. This suggestion is based on the observation that the organisms as they are gliding on agar leave behind them trails whose index of refractility is somewhat different than that of the cells (Figure 7-15). When a cell encounters a trail that has already been laid down, its rate of gliding increases considerably. There is no direct evidence, however, that the trail is composed of polysaccharide. Another possibility, in the context of the role of polysaccharide in motility, is that the slime acts as a Steffan adhesive; that is, a material that allows the cell to adhere to the substitute but does not interfere with its ability to move along the substitute.

2. It is possible that the slime surrounding the cell serves as a matrix within which excreted hydrolytic enzymes are bound. This could

Figure 7-15 Slime trails left behind cells of *Myxococcus fulvus* gliding on a thin layer of agar. Phase contrast photomicrograph. (From Reichenbach and Dworkin 1981)

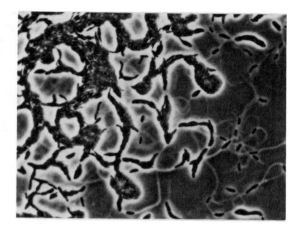

have the effect of stabilizing the enzymes, thereby increasing their longevity and thus their utility to the cell. If this were its function, one would predict that the polysaccharide would have a substantial amount of charged groups. This seems not to be the case, but it has not been sufficiently carefully examined.

3. Still another role that the slime may play is to slow down the rate of diffusion of certain molecules within the slime matrix.[18] *M. xanthus* has certain serious problems with regard to perceiving a gradient of chemoattractant. The rate of movement of the molecules, by diffusion, is orders of magnitude greater than the rate of movement of the cell. This will be discussed in more detail in the discussion of chemotaxis and motility; at this point I only wish to suggest that the ability of the cell to perceive a gradient may actually require that the diffusion coefficient of the chemoattractant be substantially reduced and this this may be accomplished by excretion of the slime.

4. The surface of the myxospore, as indicated earlier, is covered by a capsular polysaccharide comprised of α-1,3 glucan. Whether or not this is related to the other polysaccharide(s) is not clear.

5. One need only look at a fruiting body of *S. aurantiaca* (Figure 7-7 and 7-16) to recognize that there is a problem. How does a mob of prokaryotic cells organize its communal activities sufficiently to construct such an elaborate, relatively complex structure? Since the myxobacterial fruiting body contains large amounts of polysaccharide, it is likely that it plays a role both in the construction of the fruit and in its function. Since the cells of *S. aurantiaca* are continually moving through the fruiting body as it is being formed, the temporal variable can generate a spatial variable. Thus, if one thinks of the cell as continually excreting some kind of self-assembling molecules according to a temporally predetermined program, it at

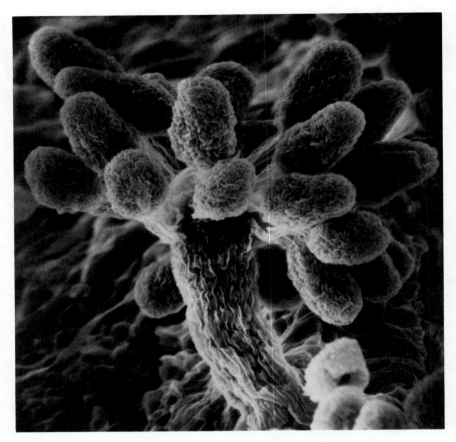

Figure 7-16 Scanning electron micrograph of a fruiting body of *Stigmatella aurantiaca*. (From White 1981)

least offers a conceptual framework within which to consider the process. One usually thinks of proteins as the self-assembling molecules. However, it has long been recognized that polysaccharides have the same property. The appropriate mixtures of polysaccharides will, under the right conditions, assemble to generate quaternary structure not characteristic of either of the components. Furthermore, the assembly can be reversible, responding either to temperature changes (as in the case with the reversible gelling and melting of agar) or to divalent cations (as in the case with alginates, responding to Ca^{++}).[19]

6. Finally, the ecological niche of the myxobacteria is not an aqueous milieu in which the cells are immersed; instead, the cells are probably partially immersed in a thin film of water on the surface of

solid particles. Thus, they are subject to the serious problem of desiccation. While the myxospore is completely resistant to desiccation (and that may, in fact, be its primary survival advantage), the vegetative cells are extremely sensitive to desiccation. The ability to excrete slime may protect the cell from desiccation in a number of ways. The slime may exist as a sheet over the cells and simply act as a water-impermeable, oxygen-permeable barrier. Or, the slime may be hygoscopic, maintaining a constant and optimal degree of residual moisture. In other words, if the slime is indeed hydrophilic, it may act as a water buffer, regulating water binding and water release.

In any case, ignorance about the nature of myxobacterial slime represents a serious barrier to understanding fruiting-body formation and morphogenesis as well as some extremely important aspects of myxobacterial biology.

G. Motility and Chemotaxis

1. Motility

The myxobacteria are included in a larger taxonomic group called the "gliding bacteria."[20] These bacteria move by gliding on a solid substrate rather than by swimming through an aqueous medium. Although the gliding bacteria are united in a single group, they include such different organisms as myxobacteria, *Thioploca, Beggiatoa, Simonsiella, Cytophaga, Vitreoscilla,* and *Leucothrix.* While it seems intuitively reasonable that movement by gliding over a solid surface may serve a common function, perhaps fundamentally different from that shared by all swimming bacteria, there is no evidence nor is it obvious that these gliding bacteria share a common mechanism of gliding or a common evolutionary pattern. Thus, this discussion of the mechanism and function of myxobacterial gliding will make no pretense of providing unitary theories; myxobacterial gliding will be discussed in its own context.

Investigations into the mechanism of gliding motility have focused on *Oscillatoria, Flexibacter,* and *Cytophaga;* only rarely and recently have the myxobacteria been the subject of these studies. Hypotheses as to the mechanism of gliding have fallen into two categories—mechanical and biophysical. The mechanical theories including spinning "rotary assemblies," moving treads, fimbrae that alternately extend and retract, and contractile fibrils. The biophysical theories include surface tension-driven movement and jet propulsion by excreted slime. Some of these hypotheses can be discarded on the basis of available evidence (e.g., retractile fimbrae, jet propulsion, or actinlike proteins); others (e.g., spinning rotary assemblies) seem unlikely. Among the remaining theories, we have proposed and subjected to analysis the idea that gliding motility in *M. xanthus* is generated by the force imbalance created by the polarized excretion of surfactant.[21] I am naturally biased in favor of this theory but must admit that, at the moment, there is no compelling reason for concluding that any of the remaining hypotheses explain gliding motility. Thus, the mechanism of gliding motility remains an enigma. However, it is encouraging that a number of laboratories are now actively investigating the problem; perhaps the next edition of this book will be able to provide some insight into this process.

In the case of swimming bacteria, it can be calculated that directed motility (but not random motility) substantially increases the flux of nutrient into the cell. Thus, a clear function of motility is to optimize feeding. This may also be the case with gliding bacteria. However, one must keep in mind that the mode of feeding of swimmers and gliders is radically different. Swimmers depend on the uptake of soluble nutrients; while in nature, gliders in general, and certainly myxobacteria specifically, feed on insoluble, macromolecular debris. Thus, the ability to glide over the surface of a particle seems adaptive and consistent with this mode of feeding. In addition, if an organism is dependent for its nutrients on an insoluble, nondiffusible substrate, motility is an indispensable attribute. A second function of motility arises from the developmental activities of the myxobacteria. Movement into aggregation centers, and in the case of *Stigmatella*, movement through the stalk into the sporangioles, obviously requires directed motility.

It is interesting that there are two modes of gliding motility in *M. xanthus*. These modes have been designated as "A" (for "adventurous") motility and "S" (for "social") motility.[22] Cells with mutations in any of the A system genes can no longer move as individual cells but only in groups of two or more closely opposed cells. Cells with mutations in any of the S system genes can move as single individual cells; moving clusters of cells are far less frequent. Given the dual and possibly multiple functions of myxobacterial motility, it is interesting to speculate on the specific functions of A and S motility. Generally, the myxobacteria move in order to find food or to aggregate as part of the developmental process of fruiting-body formation. It is possible (albeit a bit fanciful) that in each case some of these cells are acting as pathfinders, so that the movement of a swarm of cells is not committed to a particular direction until it is established that that direction is an appropriate one. Thus, A motility would be expressed by the pathfinder or scouting cells and S motility by the swarm.

While the nature of the forces that hold the cells together as they move in a swarm has not been explained, S motility is highly correlated with the presence of polar pili; furthermore, it seems that the pili function to draw cells together until they are closely opposed. The coexistence of two different and genetically separate mechanisms of motility is fascinating, both from the point of view of their different functions and the regulatory mechanisms.

2. Chemotaxis

Myxobacterial chemotaxis is a dilemma. It has never been directly demonstrated; however, its existence has been taken for granted because of two factors. First, the striking analogy between the myxobacteria and the cellular slime molds, for whom chemotaxis is a well-documented and intimately examined fact of life, often leads to the assumption that chemotaxis occurs in both groups. Second, there are a series of experiments that claim to have demonstrated chemotaxis. Some of these (including one from my own laboratory) could not be repeated. Others claim to have demonstrated chemotaxis but, in fact, only demonstrate that fruiting-body formation can be induced at certain sites. Despite heroic measures, we have been unable to demonstrate directly and compellingly that chemotaxis occurs.[23]

In retrospect, one should have expected that demonstrating chemotaxis in the myxobacteria would be a problem. In their seminal paper on flagellar chemotaxis, Macnab and Koshland pointed out that there were two general mechanisms that could describe bacterial chemotaxis. By means of a spatial perception mechanism, the cell would simultaneously measure the concentration of the putative attractant or repellant at its front and hind ends, compare the two values, and decide if it was moving in a favorable or unfavorable direction. In a temporal perception mechanism, the cell would measure the concentration at some time, store that value, move along, and then measure the concentration at some subsequent time. It would then compare the two values and determine from the comparison if it was moving to or from the chemical. Some simple calculations showed that, owing to the small size of a bacterial cell, the ability simply to discriminate between the concentrations of known attractants or repellants at the front and back of the cell involved a precision that seemed unreasonable. By means of an elegant experimental approach, it was in fact demonstrated that the perception was indeed temporal rather than spatial. For the myxobacteria there is the same problem with a spatial sensing mechanism; *M. xanthus* is only two to three times longer than *E. coli*. With regard to a temporal sensing mechanism, there is an even more interesting problem. *M. xanthus* moves slowly (about 1 to 2 μm/min; roughly 1/3000 the speed of *E. coli*). It can be calculated that the rate of diffusion of most molecules is so much greater than the rate of movement of *M. xanthus* that no matter which way it was moving with regard to the source of a molecule, it would always seem that the concentration of that molecule was increasing.

It is quite clear that directed movement of the myxobacteria does indeed take place; after all, during the process of fruiting-body formation, aggregates of cells do form. And there is now evidence that shows that M. xanthus can perceive objects such as polystyrene latex or glass beads at a distance and then move toward them.[24] However, it may be that we are going to have to be much more clever about how this happens than simply invoking classical chemotaxis.

H. Bacteriophage

The first bacteriophage isolated for the myxobacteria was designated Mx1; it was isolated on M. xanthus but could also infect other species of Myxococcus (i.e., M. fulvus and M. virescens). It is a large, virulent DNA phage, and since it was first isolated, a variety of related phages have been reported.[25] If vegetative cells of M. xanthus are induced to convert to myxospores and are infected with Mx1 before the cells undergo any substantial shape change, the phage is taken up by the inducing cell and trapped in the subsequent myxospore. Upon germination, the phage is released and the germinated cell is lysed. The mechanism of this entrapment has not been determined. The phage does not exist in the spore as an intact particle. However, neither the stage of its development that is interrupted nor the mechanism of that interruption is known.

There are three other groups of phages that have been described. All of them have been isolated with M. xanthus as the host and are small, DNA phages capable of carrying out generalized transduction. Their properties are described in Table 7-1.

The surprising discovery was made that the coliphage P1 could inject its DNA into M. xanthus.[26] Furthermore, this could be carried out with a P1 carrying the Tn5 transposon for kanamycin resistance, thus introducing a stable, selectable marker distributed randomly around the host genome.[27] This is discussed in more detail in Section I of this chapter.

The interaction between the myxobacteria and their phages is not only intrinsically interesting as a host-parasite interaction in a developing system, but it is clear that bacteriophages can be powerful probes for a variety of genetic analyses of development. Parenthetically, it is interesting and unfortunate that despite a number of attempts, no phages for other myxobacteria such as S. aurantiaca have yet been isolated. Recent work has demonstrated that Mx8 is a temperate phage,[28] and it may thus be useful for carrying out specialized transduction.

Table 7-1 Myxophage prototypes

Phage	Diameter of head[a]	Tail	Host range	DNA (kb)[d]	GC moles%	Transduction[c]
Mx1	75 mm	long, contractile	M. xanthus M. fulvus M. virescens	200	50	nt
Mx4	65 mm	long, contractile	M. xanthus M. fulvus	60[b]	70	g
Mx8	55 mm	short	M. xanthus	56	70	g
Mx9	60 mm	short or absent	M. xanthus	60	nt	g

[a] All have approximately isometric heads.
[b] Terminally redundant but not cyclically permuted.
[c] "nt" means not tested. "g" means generalized transduction.
[d] "kb"—thousand base pairs.

I. Genetics

The extraordinary growth of recombinant DNA technology over the past decade has opened up all sorts of new approaches to a genetic analysis of development. In addition, the burgeoning field of movable elements of DNA has introduced an entirely new strategy to genetic analysis. These will be explored in more detail in a subsequent chapter. At this point, it is sufficient to point out that these approaches are available in the myxobacteria. Dale Kaiser's laboratory at Stanford has pioneered the imaginative use of these new strategies, and for the first time one can confidently predict that a genetic analysis of development in myxobacteria will yield regulatory insights.

1. DNA

Two approaches have been used to determine the size of the genome of *M. xanthus* and *S. aurantiaca*: renaturation kinetics and an ingenious new approach based on a quantitative evaluation of the relative amounts of fragments of DNA generated by endonuclease treatment and separated electrophoretically.[29] While both of these approaches involved some unproven (but not unreasonable) assumptions, they both arrived at approximately the same value, 3.8×10^9 daltons. This value is about 24 to 53% greater than the generally accepted values for *E. coli* and *B. subtilis*. In fact, some uncertainties about the presence and effect of large plasmid DNA may complicate the conclusions and could result in an even lower value than that obtained.

In addition, Inouye's group noticed that a small fraction of the *M. xanthus* and *S. aurantiaca* DNA underwent rapid renaturation, suggesting the presence of repeated segments in the DNA. Recent attempts to demonstrate such segments directly have revealed the presence of 500 to 700 copies/genome of a small (163 bases), single-stranded fragment of DNA. This satellite DNA has been shown to be present both in *M. xanthus* and *S. aurantiaca*.

It seems as if there is something strange about the genome of *M. xanthus* (and perhaps about myxobacteria in general; the G + C% values of many of the myxobacteria have been determined and, despite a rather considerable variation in morphology and life cycle, all the values fall within a range of 67 to 72%). Whether this strangeness has anything to do with the developmental behavior of the myxobacteria remains to be seen.

2. Mutants

It is probably obvious that in order to be able to do a genetic analysis of development, it is necessary to have a series of stable mutants, blocked at various well-defined developmental stages. Fortunately, such mutants have been isolated, and they reflect the two general aspects of development in *M. xanthus*—myxospore formation and fruiting-body formation. With regard to the latter, it has been possible to isolate mutants (some of them temperature-sensitive) that (1) cannot aggregate, (2) are able to aggregate but do so aberrantly, (3) are able to form neither aggregates nor myxospores.[30] In another study, it was possible to show that a number of classes of fruiting body-deficient mutants could be obtained; these classes were defined on the basis of their ability to carry out a phenotypic complementation with each other. In other words, the members of one group when mixed with the members of another could jointly complete the fruiting process; alone neither could. The complementation was phenotypic in that the myxospores collected from the complementation involved the exchange of some sort of signals that the mutants could not produce. These mutants have proven extremely useful in attempts to isolate and identify the various signals exchanged among the cells during fruiting body formation (see Chapter 11).[31]

In the case of myxospore mutants, it has been possible to obtain a variety of mutants that cannot be induced (via the glycerol-induction technique; see Section J of this chapter) to form myxospores.[32] These mutants fall into a number of different classes: some cannot form myxospores under any circumstances; some can respond to one inducer (for example, phenethyl alcohol) but not to another (glycerol); most of the noninducible mutants cannot be induced but will form myxospores during normal fruiting-body formation.

M. xanthus is sensitive to the usual mutagens such as ultraviolet light, nitrosoguanidine, and ethyl methane sulfonate.[33] However, it also has a tendency to generate developmental mutants spontaneously with an extremely high frequency. For example, when cells are plated out for fruiting-body formation or when glycerol noninducing mutants are selected, the mutants appear with a frequency of 10^{-4} to 10^{-3}. The basis for this extremely high frequency of variation is unknown.

A variety of other mutants have been routinely isolated, including antibiotic resistant mutants, motility mutants, auxotrophic mutants, and others.

3. Phase Variation

The colonies of M. xanthus have a tan-orange color. This coloration is a result of their production of a red, glucosylated carotenoid called myxobacton plus a complex of undefined yellow pigments. In addition, the colonies have a characteristic medusoid quality resulting from their tendency to swarm on an agar surface. If one examines a plate full of colonies, one is sure to see a heterogeneity both of pigmentation and colony morphology.[34] Any colony when picked, subcultured, and replated is likely to give rise to the same heterogeneity. This reversible, metastable, high-frequency variability has been referred to earlier (see Chapter 2) and has been fairly well described and defined as phase variation. Phase variants of M. xanthus also seem to have different developmental capabilities. While it is possible to obtain variants that are more or less stable, the phenomenon obviously complicates genetic analyses. Neither the mechanism of the phase variation nor its biological function is understood, although the presence of a reversible, metastable, regulatory switch that can control a block of characteristics is an obviously appropriate device for developmental regulation.

4. Transduction

Three of the four groups of phages isolated on M. xanthus—Mx4,[35] Mx8, and Mx9—are capable of mediating generalized transduction. Both Mx8 and Mx9 were originally isolated from carrier strains of M. xanthus[36] and one of them, Mx8, has recently been shown to be lysogenic.[37] Transduction by Mx4 is facilitated by the use of a mutant that is temperature-sensitive for replication, has an expanded host range, and has a higher transductional frequency than the wild-type phage. Since the DNA of Mx4 is not circularly permuted nor does it insert into the host DNA, the mechanism of generalized transduction by this phage is unclear.

Mx4 and Mx9 are serologically distinct but otherwise similar. The Mx8 genome is 56,000 base pairs in length; thus, it is capable of transducing markers that are substantially separated from each other.

A variety of properties, both developmental and otherwise, have been transduced both with the Mx4 and Mx8 phages.

5. Transduction by P1

It came as a great surprise that the coliphage P1 was able to infect *M. xanthus*. While the infection was not a productive one, in terms of phage replication, P1 did insert its DNA into the host cell; in fact, a P1 carrying a Tn9 transposon for chloramphenicol resistance was able to confer chloramphenicol resistance on *M. xanthus*.[26] Subsequently, it has been possible to infect *M. xanthus* with a P1 phage carrying the Tn5 transposon for kanamycin resistance.[27] The use of Tn5 is a substantial improvement over the Tn9 chloramphenicol transposon because Tn5 in *M. xanthus* is quite stable. The use of this relatively stable transposon represented an extremely important technological advance, for it meant that it was now possible to insert a selectable marker randomly around the *M. xanthus* chromosome.

6. Use of Transposon Tn5 to Select Developmental Genes[27]

One of the problems of conventional genetic mapping of developmental genes is that the genes have no obviously selectable properties; thus, it is difficult, if not impossible, to quantitate developmental markers among recombinants. If, however, one could insert a portable, selectable marker (e.g., a transposon carrying antibiotic resistance) immediately adjacent to a developmental gene, one could then select for the antibiotic resistance with a high probability of simultaneously selecting for the developmental gene. The procedure was as follows: A developmentally competent strain was infected with phage P1::Tn5. The cells were then plated out on a medium containing kanamycin. Surviving colonies must have been all those that had received and incorporated the Tn5 transposon (bearing the kanamycin-resistance marker) in their genome. However, since the transposon inserts randomly around the host genome and since there are about 1,000 genes in the genome, one out of 1,000 surviving colonies should contain the Tn5 transposon immediately adjacent to any particular gene.

The next step then was to find this one surviving colony. The 1,000 Tn5-bearing strains of *M. xanthus* were grouped into 10 pools each containing 100 strains. Each mixed pool was infected with the generalized transducing phage Mx8. Theoretically at least one phage should package a piece of DNA containing both the developmental gene in question and the kan^r gene. The phages then were used to infect a recipient *M. xanthus* that was unable to complete fruiting-body formation. The recipients were then plated on a medium that contained kanamycin and would permit wild-type cells to form fruiting bodies. Any cell that gave rise to a clone that fruited, by formal definition, had had its developmentally deficient gene replaced by the transducing phage with a DNA fragment containing both the developmental gene and the kan^r gene. The procedure is illustrated in Figure 7-17.

This brilliant piece of work immediately made it possible to generate a library of strains containing the kan^r marker next to any previously unselectable (but screenable) mutant; it provided the *modus operandi* for mapping all of the developmental genes for which mutants could be obtained.

7. Plasmids

The search for native plasmids in *M. xanthus* has been erratic and the plasmids elusive. There have been demonstrations of low molecular weight DNA that behaved as covalently closed circles; there has been a demonstration of induction of chloramphenicol resistance at a high frequency by chloramphenicol; and there have been occasional reports of high molecular weight (10^8 daltons) extrachromosomal DNA in *M. xanthus*. None of these

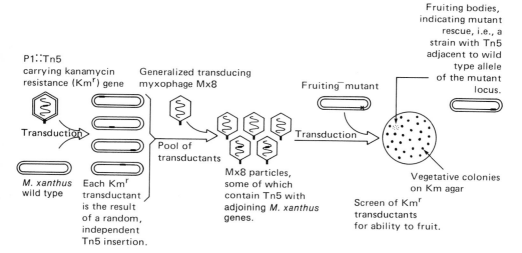

Figure 7-17 Diagram illustrating the strategy for isolating Tn5 inserted near the locus of a mutation that renders *Myxococcus xanthus* unable to form fruiting bodies. (Adapted from Kuner and Kaiser 1981)

has been subjected to really careful scrutiny; thus, at this time, it is only possible to say that it looks as if plasmids are present in M. xanthus, but the story is certainly nowhere near being pinned down.

8. Recombinant DNA Approaches

A number of developmental genes in M. xanthus have been cloned in an E. coli plasmid, methods have been devised for reinserting those genes back onto M. xanthus by P1 transduction,[38] and the stage is thus set for using this powerful approach for developmental studies.

The fact that there is no difficulty in making restriction endonuclease digests of M. xanthus DNA or in doing Southern blots of the electrophoretically separated fragments makes it possible to determine whether or not pieces of DNA are moving around during development. It is likewise possible to do restriction endonuclease mapping. While it has not yet been subjected to careful and systematic study, there is unfortunately no system available yet for transforming the cells with naked DNA.

A number of gene products have been isolated and identified as playing a role in development. In addition, there seems to be a stable mRNA made during early aggregation. This then theoretically allows the isolation of the corresponding developmental genes.

Mx8 has been shown to be a lysogenic phage for M. xanthus.[39] It can then be used as a cloning vector if, like lambda, there are indeed stretches of its DNA that are not required for replication.

Finally, recent work in Dale Kaiser's laboratory has resulted in the construction of partial diploids by means of tandem duplications. Thus, it is possible to do tests for dominance and genetic complementation. In addition, his group has constructed a fused Tn5 Lac transposon that joins LacZ expression to exogenous promoters. When transposed into M. xanthus, the transposon can now serve as a convenient indicator of promoter activity and presumably thus of gene expression (see Chapter 12).

The combination of easy availability of mutants, a variety of generalized transducing phage, a lysogenic phage, ability to put transposons into the cells, and recombinant DNA techniques—all in the context of a series of complex but easily manipulated developmental events—represent an exciting horizon in prokaryotic development.

J. Development

The peculiar aspect of myxobacterial development is that it encompasses not only a cellular morphogenesis, as do many other prokaryotes, but also a colonial morphogenesis. This latter aspect, which involves cell interactions and cell communication, is a common feature of multicellular eukaryotes but not of prokaryotes. Its occurrence in a prokaryote, especially one that has to a large extent been domesticated, offers the unusual opportunity to examine the nature of cell interactions and cell communication in an organism that is experimentally convenient and amenable to modern biochemical and genetic approaches.

When *M. xanthus* is placed on the surface of a low-nutrient medium, it will grow and divide; and when the nutrients are exhausted, if the cells are at a sufficiently high population density, they will shift from the growth mode to development and begin to aggregate. These aggregates change their shape and texture somewhat; the vegetative rods convert to round, optically refractile, resistant, metabolically quiescent myxospores, and the resultant structure is known as a fruiting body. This structure may be more (e.g., *S. aurantiaca*, Figures 7-7 and 7-16) or less (e.g., *M. fulvus*, Figure 7-5) complex. In any case, it is stable, essentially indefinitely, until physical and nutritional conditions induce myxospore germination, whereupon the whole process of swarming, feeding, and development takes place again. The requirement for a solid surface (or perhaps just an interface) needs no extensive discussion; the cells must move in order to form the aggregates. They cannot do so when in suspension.

The requirement for a high cell density is interesting. How does a bacterium know what its population density is? There are two possibilities: if the cell density requirement is high enough to require physical proximity of the cells, the signal may be a tactile one requiring some interaction between cell surfaces; alternatively, the cells may excrete a signal molecule whose external concentration they can measure and which serves a parameter of cell density. It can be shown that, in the case of *M. xanthus*, development requires that the cells be essentially contiguous. Nevertheless, it is also possible to fool a low-density population into aggregating by adding micromolar concentrations of adenine or adenosine. A number of other experiments support the idea that adenosine, or a closely related molecule, is somehow involved as a cell density signal[40] (see Figure 8-3). While the data support the notion of an extracellular signal, the requirement for contiguity of the cells leaves the nagging doubt that some sort of cell-surface interaction may also be involved.

The third factor that must be satisfied in order for cells to enter the developmental mode is the nutritional one. It had long been recognized that if myxobacteria were placed on a high-nutrient medium, they grew but did not develop. On a low-nutrient medium, however, a limited amount of growth occurred, followed by development. Based on the response of *M. xanthus* to a defined medium containing only amino acids and salts, it was originally proposed that in order for development to begin, the cells had to perceive the reduction in the concentration of two specific amino acids—

phenylalanine and tryptophan. Subsequent studies showed that development was switched on by the perception of a reduction in any of the required or growth-limiting amino acids or by starvation for a carbon-energy source or for inorganic phosphate.[41] Thus, the cell seems to be responding to a variety of nutritional circumstances which signal the end of growth. What is unclear is why cessation of growth on a high-nutrient medium is not followed by development.

Following this, some attention was paid to the question of how this perception of general nutritional deprivation was translated into the developmental response. Both of the proposals that have been made have originated by analogy with the response of E. coli to nutritional down-shift. First it was reported that cAMP stimulated fruiting-body formation. The role of cAMP as a specific messenger was made unlikely by the observation that other adenosine nucleotides such as ADP were even more effective. Next it was shown that the compounds guanosine tetra- and pentaphosphate, which are involved in the general response to starvation in E. coli, accumulate in M. xanthus under those nutritional conditions that lead to development. It was thus suggested that in M. xanthus, like E. coli, these nucleotides regulate a whole spectrum of growth-related responses that are in some way part of the switching complex that turns development on and off.[42] The isolation of a "relaxed" mutant of M. xanthus (that could not respond to starvation with elevated levels of ppGpp) would help to prove or disprove this hypothesis.

1. The Sequence of Developmental Events

If the above three requirements are satisfied, cells will initiate development, and the sequence of events that follows is illustrated by Figure 7–18. If a heavy suspension of cells is evenly spread over the surface of an appropriate agar medium, one can see the beginning of aggregates at about 15 h. These are referred to as *preaggregates*, and at about this time, massive lysis (80 to 90%) of the population occurs. At about 30 hours, the aggregation process per se is completed and these are referred to as *aggregates*. At about 35 hours, the surviving vegetative cells begin to convert myxospores and the structures are now referred to as "immature fruiting bodies." The structure continues development until about 72 hours, becoming increasingly darkened and well defined, at which point it is referred to as a *mature fruiting body*.

2. Marker Events during Development

One of the problems that has continually beset studies of development is that developmental events are usually described in morphological rather

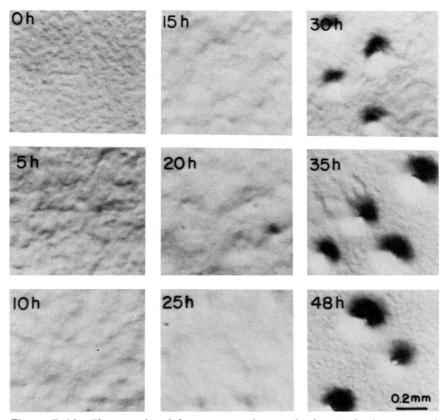

Figure 7-18 Photographs of the sequence of events leading to the formation of fruiting bodies by *Myxococcus xanthus*. At 0 time the vegetative cells are evenly spread on the surface of a solid medium designed to support fruiting body formation. (From Orndorff 1981)

than biochemical terms. Fortunately, a number of events occur during development that can be defined more precisely and thus can serve as biochemical markers of developmental events; some of them may also play a causal role in the developmental events.

One approach has been to examine the protein composition of cells during development, via acrylamide gel electrophoresis. When this examination was done with *M. xanthus*, it was shown that about 25% of the 30 major soluble proteins that were resolved underwent significant changes during development. The authors grouped these changing proteins into three classes—accumulation proteins, peak proteins, and late proteins.[43] This study showed (not surprisingly) that morphological development was associated in an obviously complicated way with changing patterns of protein synthesis. It holds open the promise that specific protein changes can be associated with specific morphological events.

One such protein has been isolated, rather thoroughly characterized,[44] and shown to play a structural role in the mature myxospore. This protein, called *protein S*, is synthesized early in fruiting-body formation and at that time can be isolated as a soluble component of broken vegetative cells. Its peak-rate synthesis represents 15% of that of the total protein synthesis during that period. Later on in development it becomes insoluble and a self-assembling outer component of the spore coat.[45]

Another development-specific protein called "myxobacterial hemagglutinin" (MBHA) has been demonstrated, isolated, and thoroughly characterized. MBHA is a lectin with hemagglutinating activity. It is present only in cells undergoing fruiting-body formation and reaches its peak in cells at the point of maximal aggregation. While it has not been possible to demonstrate that it plays any role in aggregation, its properties and the timing of its appearance would suggest that this is indeed the case.[46]

It has been assumed, so far without any direct supporting evidence, that cell-surface interactions will turn out to play a role in the cell interactions involved in myxobacterial development. With this in mind, the proteins of isolated outer membranes of *M. xanthus* have been separated and followed during development.[47] The rates of synthesis of a number of them have been shown to change drastically during aggregation; however, studies with mutants have shown that none of these changes are necessary for aggregation per se. Presumably, then, they are involved in some late stage of fruiting-body formation—perhaps spore formation. In this context, it is interesting that S protein (neither S protein nor MBHA, by the way, is one of the outer membrane proteins referred to earlier) is synthesized during early aggregation even though it becomes functional only during late development.

A somewhat less well-defined marker of developmental events is represented by RNA synthesis during fruiting-body formation. First, a method had to be devised to measure rates of synthesis during development on a solid surface. The method was a variation of one used to study macromolecular synthesis during development in *Dictyostelium discoideum* and was based on the fact that cells will undergo development on the surface of a Millipore, Unipore, Nucleopore, or similar filter. Thus, the cells on the filter are pulsed with a radioactive precursor of the macromolecule being studied by placing the filter and the cells atop a pile of adsorbant filter pads soaked with the labeled precursor. After the desired pulse period, they could be removed and placed in the presence of an inhibitor, or unlabeled precursor, to chase out the label; the filter could be placed back on the surface of the medium and allowed to continue development, or the cells collected and the incorporation of precursor determined. In all such studies, when attempting to measure rates of synthesis, it is critical that one demonstrate that the change in the rate of biosynthesis is not attributable to a change during the measurement process; that is, one must show that the intracellular specific activity of the precursor remained constant. Using this approach, it was possible to measure a characteristic pattern of changes in rates of RNA synthesis during fruiting-body formation.[48] During early aggregation, the cells synthesized a fraction of RNA that sediments between 5S and 16S and, on the basis of a number of criteria, seemed to be a fraction of mRNA that is substantially more stable than that formed by vegetative cells. It has recently been shown that this stable mRNA is probably coding for S protein.[49]

The issue of biochemical correlates of development will be dwelt upon in greater detail in a later chapter. At this point I just want to emphasize that in order for developmental events to be characterized and understood in terms of the regulatory processes that control them, it is a *sine qua non* that these events be defined or at least marked biochemically.

3. The Function of the Myxobacterial Life Cycle

The question as to the function of this rather elaborate life cycle naturally arises. I have proposed that its role can be understood in the context of the peculiar feeding habits of the myxobacteria.[50] All of them feed on macromolecules of one sort or another—protein, polysaccharide, nucleid acids— as well as on insoluble organic debris such as bacterial carcasses. In addition, they are able to lyse and feed on living bacteria and do so by excreting a battery of potent hydrolytic enzymes. It seems obvious that for such an organism it is a distinct advantage to concentrate these extracullular hydrolytic activities by existing at all times in as high a population density as possible. A single cell excreting hydrolytic enzymes will not be able to achieve and maintain a sufficiently high concentration of these enzymes to be able in turn to maintain a growth-supporting level of diffusible, low molecular weight substrates from the target particle. On the other hand, a high-density population of cells will be more able to do so. In other words, the cells manifest a kind of wolf-pack effect.

In fact, this prediction is experimentally borne out. There is a clear and substantial effect of cell density on the growth rate of *M. xanthus* with casein as a substrate. When enzymatically hydrolyzed casein (e.g. Casitone) is used instead, there is no such cell-density effect.[51] Thus, the life cycle of the myxobacteria seems to function to maintain at all times a sufficiently high cell density to allow the cells to feed efficiently as a swarm. The function of the aggregation process culminating in a fruiting body may then be to regroup the population that has been dispersed during feeding and is about to enter a resting state; upon germination of the spores, when conditions are once again appropriate for growth, the cells are already present as a high-density population.

4. Spore Formation

The myxospore of M. xanthus differs from endospores in that the entire cell converts to the spore by shortening and rounding up, unlike the endospore which forms within a portion of the vegetative cell. This difference is not trivial. It has been demonstrated that during endospore formation there is a considerable amount of macromolecular turnover; actually, this is not so much turnover as it is an exchange of materials between the coexisting cells — the vegetative mother cell and the developing spore. There is now good evidence to indicate that at least for a period of time during sporulation, the two two genomes in Bacillus subtilis are functional and are both participating in the sporulation process.[52] In myxospore formation, since the entire vegetative cell converts to the spore, there is no such nurturing or cross-feeding of the developing spore by the mother cell. The suggestion has been made that the function of the massive lysis that occurs during aggregation is to serve this function of cross-feeding. However, instead of part of the cell lysing so that the remainder of the cell can complete its development, part of the population lyses to allow the remaining 10 to 20% to complete its development.[52] It is also formally possible (and even more intriguing) that the lysing population contributes an extracellular, structural component of the myxspore (protein S?). In a sense, developmental autolysis may be an interesting example of altruism at the microbial level.

I made the observation that microbiologists are uncomfortable studying processes that do not take place in liquid suspension. Progress of research on myxobacterial development was materially improved by the discovery that it was possible to induce vegetative cells of M. xanthus to convert to myxospores in liquid suspension without going through the ordinary life cycle. Thus, one could short-circuit the process of normal development by adding one of a number of inducers to vegetative cells growing exponentially in a complex medium. The cells responded by converting to myxospores rapidly (in about 100 to 120 min), synchronously and relatively completely (95%).[13] The sequence of morphological changes is illustrated in Figure 7-19. Glycerol at a relatively high concentration (0.5 M) was the most effective inducer; little if anything is known about the mechanism whereby glycerol turns on this particular morphogenetic switch (although it has been shown that in E. coli, equivalent concentrations of glycerol appear to disrupt the process of methylation of the methyl-accepting proteins involved in regulating the chemotactic response.)[54]

There has been some debate in the literature as to the extent to which glycerol-induced myxospores differ from fruiting body spores. This matter is obviously of some importance, for if artificially induced spores are to be used as a model system with which to examine the properties of myxospores as well as the morphogenetic mechanisms leading to their development, one must have a fairly clear notion of where they are similar and where they

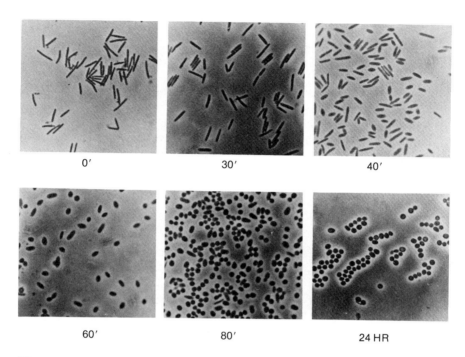

Figure 7-19 Morphological changes during glycerol-induced conversion of vegetative rods of *Myxococcus xanthus* to myxospores. Phase contrast photomicrograph. (From White 1975)

differ. From a morphological point of view, the events leading up to the spore are essentially similar in both types of spores; however, the outer surface of the two spore types are rather different. Fruiting-body spores contain a thick spore coat with S protein as the outer layer. The S protein can be removed by high salt and can then be replated onto the spores. Glycerol spores have a relatively thin spore coat, no S protein, and cannot, in fact, be replated with isolated S protein. In addition, glycerol spores lack the MBHA associated with fruiting-body spores. From a physiological point of view, glycerol spores have a higher rate of endogenous metabolism, can be germinated more easily in distilled water, and are somewhat less resistant to disruption than are fruiting-body spores.

As long as these differences are kept in mind, they are not critical differences, and the glycerol-induction method may be used quite effectively for asking questions about the regulation of spore induction, the morphogenetic processes leading to shape change, and the acquisition of resistance.

One final note on properties of *M. xanthus* myxospores: When the adenylate energy charge of glycerol spores was determined, it was found to be 0.84, comparable to 0.81 for vegetative cells. In view of the low energy charge for *Bacillus* endospores (0.1), the premature conclusion was drawn that when bona fide fruiting body spores would be examined, their energy charge would be low like that of endospores, indicating yet another difference between glycerol and fruiting-body spores. However, we were surprised to find that the energy charge of fruiting-body spores was relatively high (0.73) and not significantly different from that of vegetative cells.[55]

Vegetative cells of *S. aurantiaca* convert to myxospores during fruiting-body formation in that organism. The spores are somewhat shortened, have a thin spore coat, and are more resistant than vegetative cells to desiccation and sonic oscillation. Like *M. xanthus*, the spores can be artificially induced in liquid suspension. Unlike *M. xanthus*, there is a wide variety of treatments and inducers that will effect the conversion.[56]

5. Relationship between Myxospore Development and Fruiting-Body Formation

Is there a causal relationship between fruiting-body formation and myxospore formation? For example, does the fruiting body or the aggregate create an environment that induces myxospore formation? One could imagine a scenario such as the following. The aggregated cells accumulate metabolic CO_2 which is trapped by the slime formed by the aggregate. The CO_2 or a low pH turns on myxospore formation. Again, the ability to isolate mutants made it possible to distinguish between different models. Mutants that could sporulate but could not aggregate have long been available.[57] However, a detailed analysis of a large number of mutants that were temperature-sensitive for various stages of aggregation made it quite clear that there are indeed two separate and essentially independent pathways—one for aggregation and another for sporulation.[58] Nevertheless, they are not completely independent, as some mutants were blocked both in sporulation and aggregation. The nature of the regulatory interaction between the two processes remains unclear.

6. Spore Germination

Under the appropriate conditions, the myxospores of *M. xanthus* will germinate as illustrated in Figure 7-20. The capsule and spore coat are lysed, and the cell emerges and gradually elongates. (Most of the work that has been done on myxospore germination has used glycerol-induced myxospores. Keep in mind the caveat expressed earlier—while these are in

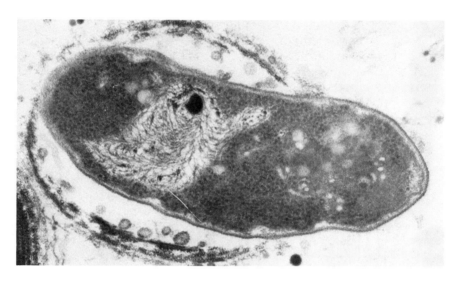

Figure 7-20 Electron micrograph of a thin section of a myxospore of *Myxococcus xanthus* during an advanced stage of germination. Magnification is 88,000×. (From Voelz 1966)

fact myxospores, there are some important differences between them and fruiting body spores.)

If a suspension of glycerol-induced myxospores is placed in a growth medium containing a concentration of organic and inorganic nutrient that will support optimal growth, the spores germinate in about 30 to 40 min. Not surprisingly, synthesis of RNA, DNA, and protein takes place during this period and seem to be required for germination to occur. There is no detectable heat activation. If, however, the spores are placed in distilled water rather than growth medium, an interesting phenomenon is observed. Under these conditions, the spores will germinate only if they are present at a sufficiently high cell density, about 10^9/ml. This immediately raises the same interesting question as was posed earlier; namely, how a spore knows what the cell density of its population is. The spore uses the same simple and cunning strategy as do the cells when preparing to fruit. They excrete orthophosphate when suspended in distilled water and then measure the external concentration of the phosphate. The concentration is then a parameter of cell density and when it is sufficiently high, the spore germinates.[59] The function of this behavior is likewise interesting and consistent with the notion of the overall function of the myxobacterial life cycle. As described earlier, the life cycle seems to function to maintain a high cell density or the potential for such. Thus, the simultaneous germination of a high density of spores guarantees the immediate formation of the feeding-efficient swarm.

7. Extracellular Complementation of Development

A few years ago, Dale Kaiser's laboratory at Stanford made the extremely important discovery that it was possible to isolate mutants of *M. xanthus* that were able to repair each other's developmental deficiencies. In other words, a mutant in class A was unable to complete its development (i.e., form fruiting bodies or myxospores); likewise for a mutant in class B. However, if the mutants of the two different classes were mixed together, they could complete development. The complementation was not a genetic one; the complemented cells retained their developmental mutation. The model that emerged from these observations was that each mutation represented an inability to generate a particular chemical signal and that the complementation reflected the exchange of these signals by different classes of mutants. Altogether, four complementation groups were found, suggesting the existence of at least four signals whose production (and exchange in a normal population?) was necessary for the completion of development. In a similar fashion, parental wild-type cells could complement all four complementation groups.[31]

This work was important for at least two reasons. First, it suggested that at least four factors were produced that could be exchanged between cells and that were necessary for the completion of development. It was tempting to think that these represented four intercellular signals that functioned in some way to coordinate the developmental activities of the population. Second, it provided an excellent experimental system for attempting to isolate these signals. For if one has a mutant that is unable to complete development but can do so if a factor is added, one has, of course, an excellent assay for the factor. At the time of this writing, more than one laboratory is actively pursuing the isolation and identification of these factors. Furthermore, the wild-type genes that are mutated in a number of the conditional mutants have been cloned; thus, the possibility of determining their gene products is in the near future.

K. Stigmatella aurantiaca

It is a common joke among myxobacteriologists that when giving a seminar on the myxobacteria, one shows pictures of *Stigmatella* and then proceeds to describe the work of *M. xanthus*. Among other things, this jest testifies to the beauty and complexity of the *Stigmatella* fruiting body (Figures 7-7 and 7-8).

There are a number of reasons why one might choose to work on *Stigmatella*. The fruiting body of *S. aurantiaca* is among the most complex of the myxobacteria. It thus offers the opportunity to ask questions about the forces and factors that allow a simple prokaryotic cell to construct such an elaborate, regular macrostructure (more about that later). Second, it is the only nonphotosynthetic prokaryote that shows a developmental response to visible light.[60] Third, it has been possible to examine in some detail a few interesting developmental interactions. All of the work on *S. aurantiaca* has been done in David White's laboratory at Indiana University.[61]

The life cycle of *S. aurantiaca* is illustrated in Figure 7-4. The cells grow dispersed in liquid culture and will undergo relatively synchronous development when placed at a sufficiently high cell density on the surface of the appropriate low-nutrient agar surface. In order for optimal fruiting-body formation to take place, the cells must be incubated in the presence of visible light; the blue portion of the spectrum (410 to 470 nm) was found to be the most efficient, however, the nature of the photoreceptor(s) is unknown.

It is not clear what factors are involved in the aggregation process that is a prelude to fruiting-body formation. As indicated earlier, all of us have assumed by analogy with *Dictyostelium* that chemotaxis was the central part of the process. However, my own thinking now is that the communication may be more akin to shaking hands than to shouting across relatively large distances. In other words, communication may be largely via tactile, surface interactions than by diffusible, extracellular signals.[62] *S. aurantiaca* has offered a nice experimental system for pursuing this approach. If cells that have begun development on a solid surface are removed and placed in liquid suspension, they rapidly form multicellular clumps and settle out of suspension. Such clumping reflects an increase in the cell adhesiveness during development and is specific in that the clumping only takes place between similar cell types (other myxobacteria are excluded). This behavior of the cells offers the opportunity to ask questions about the nature of cell surface changes that may be causally related to aggregation.

Another interesting aspect of development in *S. aurantiaca* is that it has been possible to isolate a relatively low molecular weight compound that is excreted by developing cells and that has some clear-cut effects on development. When *S. aurantiaca* is present at a low cell density, under conditions that would otherwise allow development, the process is slow, inefficient, and has an absolute requirement for visible light. In the presence of the isolated factor, the time required for development is drastically reduced, the number of fruiting bodies formed is increased, and the requirement for light is abolished. White and his coworkers refer to the factor as a "pheromone."[63] While that is a technically accurate use of the term it has not yet been inequivocally established that the factor is an extracellular attractant. In any case, it is an interesting and useful experimental handle on the problem of chemical signaling in myxobacteria.

1. Fruiting-Body Morphogenesis

The structural complexity of the fruiting body is a challenge to anyone who has ever pondered the question of the organization of multicellular structure. The ability of these simple prokaryotic cells to construct such a complex, regular multicellular object is baffling and is one aspect of myxobacterial development that no one is working on. In Chapter 9, I shall suggest an idea that offers an experimental approach to the problem.

L. Conclusions and Salient Questions

From a behavioral and developmental point of view, the myxobacteria are the most complex of the prokaryotes. Yet they have been sufficiently domesticated so that they can be handled and manipulated with relative ease in the laboratory. A great deal of fundamental information has been obtained concerning their structure, growth, and metabolism; many of their idiosyncrasies have been tamed, and the stage is now set for the emerging genetic and molecular analysis of their developmental behavior. The rare opportunity to examine cell interactions and the various kinds of signaling processes involved in cell to cell communication in a prokaryotic cell is an exciting prospect.

The larger salient questions concern the molecular details of the initiation and regulation of development:

1. What are the intracellular signals whereby the external environment controls the initiation of development?
2. What are the genetic regulatory mechanisms that control gene expression?
3. What is the precise nature of the cell to cell signaling that occurs among the myxobacteria?
4. What are the signals? (This group includes both tactile, cell-surface signals, and extracellular, diffusible signals.)
5. What are the receptors?
6. How is signal reception transduced into a behavioral response?

7. At the molecular level, how are the various developmental events regulated?
8. What is the mechanism of gliding motility?
9. What is the basis of the cells' directed movement?
10. What is the function and genetic basis of myxobacterial phase variation?
11. What is the mechanism of fruiting-body morphogenesis?

PART II

THE PROBLEMS AND THE ISSUES

Chapter 8

The Relationships between Environment and Development

A. Introduction

It is possible to describe the relationships between the environment and developmental change in a phenomenological and descriptive way. Thus, when organism X encounters condition A, it does such-and-such. What I would like to do, however, is to try to provide a biological rationale for microbial development. That involves trying to understand what it is about the developmental change that makes it a biologically useful response to an environmental circumstance. Thus, it is critical that we understand the functions of the various developmental organelles or processes; this is not difficult in some cases—the function of an endospore or a myxospore seems intuitively obvious. However, the function of a *Caulobacter* stalk or of a myxobacterial fruiting body is not so obvious. So in some cases the discussion will be on reasonably firm ground; in other cases it will be speculative.

In bacteria, as in higher organisms, there are a number of different cell types that are referred to as spores. They can be resistant, resting cells (e.g., endospores and myxospores). They can be a dispersal mechanism (e.g., the motile zoospores of the *Actinoplanes*),[1] or they can be a reproductive device (such as the baeocytes of pleurocapsalean cyanobacteria).[2] These spores have obviously different functions, and it is not even clear that they should all be called *spores*. What I shall do then is to categorize a variety of developmental changes in terms of their function and then attempt to rationalize each of them considering the environmental signal to which it responds. Where possible, this categorization will be accompanied by some attempt to present a biochemical description of the process.

1. Formation of a resistant, metabolically quiescent resting cell, e.g., endospores and myxospores.
2. Formation of a dispersal mechanism, e.g., zoospores and flagellated swarmer cells.
3. Formation of a reproductive cell, e.g., baeocytes.
4. Formation of metabolically specialized cells, e.g., cyanobacterial heterocysts.
5. Formation of an optimally efficient feeding cell, e.g., the stalked cell of *Caulobacter*.
6. Formation of an aggregation device, e.g., myxobacterial fruiting bodies.

B. Formation of Resistant, Metabolically Quiescent Resting Cells

Cells such as endospores and myxospores are formed in response to an environmental situation that signals hard times from a nutritional point of view.[3] The cell that responds in this fashion has an interesting and delicate balancing act to carry out. On the one hand, since it cannot predict starvation conditions but can only experience them, the cell must respond at a point when it is fairly obvious that growth can no longer occur at any useful rate but while macromolecular synthesis necessary for development can still take place. Endospores of *Bacillus* have three separate but integrated responses. (1) They begin to accumulate carbon sources that they can draw upon during sporulation. (2) Essential growth reactions are switched off. (3) They begin a massive macromolecular turnover of proteins and nucleic acids so that, in effect, part of the pre-existing vegetative cell is used to construct the spore.

In the laboratory one can induce cells to form endospores generally in one of two ways. Either they are allowed to deplete the nutrients in a growth medium and at the postexponential stage of growth they begin to sporulate; or, alternatively, they are removed during exponential growth and resuspended in another medium that would ordinarily support growth at a very reduced rate. Under these "replacement" conditions they will then sporulate. There is no single unique substance whose absence or depletion turns sporulation on. Generally, however, the absence of an amino acid or, more effectively, of a group of amino acids will induce sporulation.

Investigators have been intrigued by the problem of how this depletion is sensed and then used to set the sporulation events in motion. There have been three general hypotheses proposed: The first was derived by analogy with *E. coli* and was based on the notion of control by catabolite repression (see Chapter 3, Section G). Since it could be shown that glucose could suppress sporulation, it was suggested that it did so in the same fashion whereby glucose or other catabolites could suppress induction of β-galactosidase.[4] It was subsequently shown that this suppression in *Escherichia coli* was mediated by cyclic AMP, whose level responded to the presence or absence of glucose and which acted as a positive control for transcriptional expression of the gene. Unfortunately, cAMP was not found in two species of *Bacillus* where it was looked for carefully (*B. licheniformis* and *B. megaterium*) even though these organisms showed glucose repression of sporulation.[5] At the moment there are two remaining theories to explain initiation of sporulation. One of these, like catabolite repression, originates by analogy with *E. coli*. It suggests that sporulation is under "stringent" control. That is, a whole set of functions are controlled in some unknown fashion by guanosine tetraphosphate (ppGpp) or guanosine pentaphosphate (pppGpp). In *E. coli*, the synthesis of those nucleotides is turned on by a nutritional shift-down, for example, when cells are shifted from a rich nutrient broth to a defined, glucose-salts medium. The nucleotides then turn off a whole series of processes, including the synthesis of ribosomal RNA and protein. The result of this is a cessation or reduction in growth. There is evidence both for and against the applicability of this model to sporulation in *Bacillus*. In favor of the model, it is possible to demonstrate the presence of (p)ppGpp in *Bacillus*, and during induction of sporulation, their levels respond to amino acid deprivation analogous to the stringent response in *E. coli*. On the other hand, a "relaxed" mutant of *B. subtilis* (a "relaxed" mutant cannot form (p)ppGpp in response to amino acid deprivation) will not sporulate unless the synthesis of GTP and GDP are prevented by inhibitors. Thus, it seems that the reduction in the levels of GTP and GDP synthesis is the key metabolic event that actually turns on the sporulation process rather than the actual levels of (p)ppGpp.[6]

An alternative model has been proposed that postulates that a different set of intracellular signaling molecules are involved. It is proposed that the cell responds to nutritional deprivation by synthesizing one or more nucleotides that function to turn on sporulation and turn off growth. These compounds are called "highly phosphorylated nucleotides" (HPN) and include such compounds as p_3Ap_3 and p_2Ap_2. When cells sporulate as a result of being deprived of certain carbon sources, p_3Ap_3 is produced. A mutant that cannot sporulate owning to a block at the initial stage of sporulation is unable to synthesize p_3Ap_3; spontaneous revertants (presumably single-site revertants) simultaneously regain the ability to sporulate and to synthesize p_3Ap_3.[7] The role that HPN play in regulating sporulation is a hotly disputed issue.

The myxobacterial spore (the myxospore) has not been examined with the same detailed scrutiny as has the *Bacillus* endospore. However, one interesting similarity with the *Bacillus* endospore has emerged. During

Chapter 8 The Relationships between Environment and Development

starvation under conditions leading to fruiting-body formation, the cells accumulate substantial amounts of guanosine tetra- and pentaphosphate.[8] While development under these conditions leads eventually to myxospore formation, the technical inconvenience of separating myxospore induction occurring within the fruiting body from induction of the fruiting bodies themselves made it difficult to attribute the nucleotide accumulation specifically, either to sporulation or to fruiting-body formation. Obviously, the use of mutants would represent a critical approach to the problem. The ideal mutant would be a relaxed mutant, unable to form guanosine tetra- or pentaphosphate under the starvation conditions that ordinarily lead to development. This would certainly support the thesis that these nucleotides are a regulatory link between the nutritional milieu and development; it would also indicate whether fruiting body formation, sporulation, or both are responsive to the nucleotide signal. It is relevant, however, that guanosine tetra- and pentaphosphate accumulate transiently during glycerol induction of myxospores.

Initially, it was felt that the glycerol induction system for myxospore formation (Chapter 3, Section J-4) offered an excellent opportunity to examine the early events that switched the cells from a vegetative to a myxospore mode. While it was quite obvious that neither 0.5 M glycerol nor any of the other inducers such as dimethylsulfoxide or ethylene glycol were the natural inducers, there was the optimistic hope that examining the process

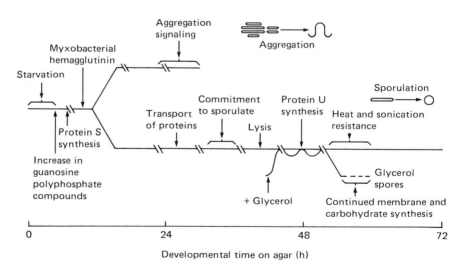

Figure 8-1 Model describing the relationship between the developmental events of aggregation and myxosporulation in *Myxococcus xanthus*. Both pathways are triggered by nutritional signals; the glycerol induction, however, can bypass the nutritional signal and induce sporulation directly. The double bar interruptions indicate sites in the pathway for which there are mutants. (Adapted from Zusman 1982)

would provide some clues as to the early triggering events. The picture that seems to be emerging now, however, is that glycerol probably short-circuits the early induction events and induces sporulation by turning on some step downstream from the initial cell-environment interaction. A question that emerges from these considerations is whether myxosporulation is a series of events that exist in a linear, dependent relationship with earlier events of aggregation. Alternatively, it may be a parallel pathway sharing some common early steps with aggregation and fruiting. Examination of a large number of developmental mutants has favored the latter interpretation.[9] This view is illustrated by Figure 8-1, which indicates the proposed relationships among fruiting-body formation and myxosporulation in *M. xanthus*. A more compelling approach to this question will involve the use of other mutant analyses that have been devised for distinguishing linear, dependent relationships from parallel, connected paths. These analyses, which will be discussed in Chapter 12, are the use of epistatic mutants and temperature-shift experiments with hot and cold temperature-sensitive mutants.

C. Formation of Dispersal Mechanisms

There has been relatively little work done on the environmental stimuli in bacteria that turn on those developmental processes leading to dispersal. Probably the two systems that exemplify the process are the release of motile zoospores from the sporangium of *Actinoplanes* (see Chapter 6) and the formation of a flagellated swarmer cell in *Caulobacter* (see Chapter 4).

Actinoplanes, like *Streptomyces*, are members of the actinomycetes. Their life cycle parallels that of the aquatic fungi in a remarkable fashion. The organism forms a sporangium, a saclike structure packed with spores. When the sporangium, which is quite hydrophobic, is thoroughly wet, the spores swell, become flagellated, and burst out of the sporangium. This process, called *dehiscence,* presumably functions as a dispersal mechanism. One sees this process of alternating phases of aggregation and dispersal throughout the biological world. Whether it is the myxobacteria and the slime molds alternately aggregating and swarming or the *Actinoplanes* alternating sessile hyphal growth with motile zoospore dispersion or insects alternating a communal and a dispersal phase, it is a strategy played out over and over again. A variation on this theme is manifested by *Caulobacter* where the sessile stage is a unicellular one but does alternate with the swarmer stage. One may view the flagellated swarmer cell (which does not grow, divide, or synthesize DNA) as a cell whose function is to maximize exploration and the stalked cell as a cell whose function is to attach to a favorable substrate and produce swarmer cells; then the fact that carbon limitation (in chemostat experiments) inhibits stalk initiation and prolongs motility makes perfect sense. The nature of the regulatory link between the environmental signal and the developmental events is unknown.

D. Formation of Reproductive Cells

There are simply no biochemical or otherwise analytic data that allow any mechanistic conclusions regarding specialized prokaryote reproductive cells.

E. Formation of Metabolically Specialized Cells

In this category, the system that has received the most detailed attention and that is most clearly defined is the formation of heterocysts by the filamentous cyanobacteria. As indicated in Chapter 5, this is a bona fide differentiation and, as such, it is a response to a different set of environmental signals than is represented by sporulation. It is now widely accepted that in the heterocystous cyanobacteria the heterocyst is the principal site of fixation of dinitrogen. Thus, the observation made many years ago[10] that removal of fixed nitrogen induced heterocyst formation now makes perfect sense from a biological point of view. The sequence of events following the removal of ammonium ions from a culture of *Anabaena* is illustrated in Figure 8-2. Proheterocysts, the precursor cells to the mature heterocysts, begin to appear within about three hours and are gradually converted to the heterocysts. The mature heterocysts begin to appear at about 16 hours and their appearance is concomitant with the appearance of nitrogen fixation. The process of proheterocyst formation in *Anabaena cylindrica* can be reversed until about two to three hours after the removal of ammonium ions by the readdition of ammonium. The conversion of proheterocysts to mature heterocysts can also be reversed during an additional two or so hours by the readdition of ammonium. In addition, removal of fixed nitrogen has to be preceded by a period of incubation in the light. The longer this period of induction—up to nine hours—the greater a percentage of cells in the filament that could convert to heterocysts.

This developmental response makes sense; it is obviously useful for the organism to be able to develop a dinitrogen-fixing alternative when the supply of fixed nitrogen is exhausted. Likewise, the absence of the oxygen-generating photosystem II in heterocysts is rationalized by the O_2 sensitivity of nitrogenase. Finally, the spacing of the heterocysts probably results in an

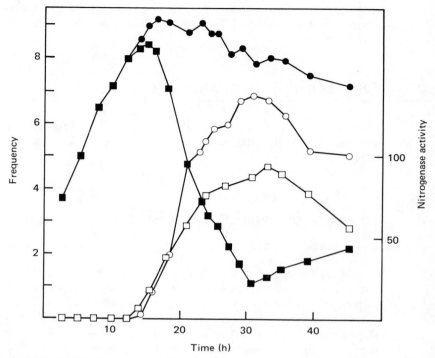

Figure 8-2 The temporal relationships among nitrogen starvation, nitrogen fixation, and heterocyst development in *Anabaena cylindrica*. ●, Total differentiated cell frequency; ■, proheterocyst frequency; □, mature heterocyst frequency; ○, nitrogenase activity as a percentage of the final activity. (Adapted from Carr 1979)

optimum ratio of cells that are generating reducing power (the vegetative cells) and those that are consuming it for dinitrogen fixation (the heterocysts).

Thus, it is fair to say that we understand the broad outline of the biological function of heterocyst development; it also seems quite clear that glutamine plays a key regulatory role in heterocyst formation. What remains to be determined is the precise nature of how glutamine is coupled to the regulation of heterocyst morphogenesis.

F. Formation of Optimally Efficient Feeding Cells

Part of the function of *Caulobacter* development (referring to the swarmer cell) has already been discussed in Section C of this chapter. It seems intuitively reasonable that the swarmer cells serve a dispersal function; and although chemotaxis has not been demonstrated in *Caulobacter*, given its inclination to flourish under conditions of extremely limited nutrients, it is very likely that the swarmer cell is, indeed, chemotactic. Having found a source of nutrient or an otherwise appropriate site, the swarmer cell drops its flagellum, develops a stalk, and settles down. It is likely that stalk formation accomplishes at least one and perhaps all of the following: (1) The increased surface area must be of considerable value to a cell seeking to maximize uptake of nutrients at a low concentration. (2) If a source of nutrient is found, the ability to attach to it or near it via the stalk would certainly be useful. (3) Since *Caulobacter* is a strictly aerobic organism, the possibility that the stalk acts as a flotation device to keep the cells near the water-air interface is a reasonable one.

Chemostat studies of synchronized populations referred to earlier indicate that carbon limitation postpones stalk formation and that stalk elongation is promoted by low levels of phosphate. Thus, evidence is available to support the link between maximizing the efficiency of food gathering and development.

Again, the mechanism whereby the cell is sensing this environmental situation is unclear; while there is ample precedent for mechanisms whereby cells sense and respond to levels of nutrient and oxygen, the lack of biochemical and physiological information in *Caulobacter* makes it difficult to understand the nature of the regulatory couple between the nutritional signal and the developmental response.

G. Formation of Fruiting Bodies

I suggested earlier that a principal function of the myxobacterial life cycle was to ensure that feeding vegetative cells were always present at a high cell density. This function not only rationalizes the tendency of cells to move en masse but also suggests a rationale for the fruiting body. Since the myxospores are usually formed within the fruiting body, the fact that they are packed tightly together in the fruiting body guarantees that, upon germination, the vegetative cells will exist immediately as a swarm. Even though the mutant analyses referred to earlier (p. 155) indicate that the pathways of sporulation and fruiting body formation are parallel, they do seem to share some common early steps. Thus, one may ask about the regulation of those early common events with reasonable confidence that one is dealing with both later developmental manifestations.

I have already pointed out that for development to take place, three general requirements must be satisfied. First, the cells must see a nutritional shift-down; this process has already been discussed in the context of guanosine tetra- and pentaphosphate. Second, the cells must be on a solid surface so they can glide. Two lines of evidence suggest that the ability to glide is necessary for development; (1) in time-lapse motion picture films of fruiting-body formation, one can see streams of cells moving into aggregation centers; and (2) more compelling, single-site, nonmotile mutants are unable to form aggregates and fruiting bodies. It is, however, formally possible that the cell perceives that it is on a solid surface, and it is that perception, rather than the movement per se, that is part of the cascade of events that sets aggregation in motion. The third requirement for aggregation is that the cells be present at a sufficiently high cell density (Figure 8-3). The mechanism whereby this is perceived has already been discussed (Section J of Chapter 7), and the data thus far indicate that one strategy that the cells use is to use the concentration of some excreted molecule as a parameter of cell density. Another possible strategy is that the cells must be in physical contact with each other. This strategy will be discussed in Chapter 11.

Finally, any discussion that attempts to clarify the relationship between environmental circumstances and developmental change must address itself

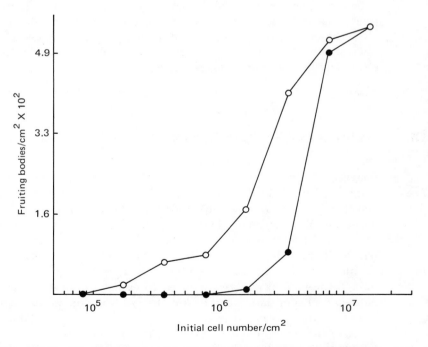

Figure 8-3 The dependence of fruiting body development on cell density in *Myxococcus xanthus*. In the presence of 50 μm adenine (O), fruiting body formation occurs at lower cell densities than in the absence of adenine (●). (Adapted from Shimkets and Dworkin 1981)

to the question of the function of myxobacterial fruiting bodies. I have already suggested that it is useful, if not obligatory, that the spores that form within the fruiting body be present at a high cell density so as to optimize swarm movement and collective feeding upon spore germination. But if this were the sole function of the fruiting body, how does one understand the elaborate complexity of some of the structures? The division of the aggregate into cysts may represent some optimization of swarm development and dispersion in that each cyst may contain an optimal number of cells with which to initiate a swarm; individual cysts could then represent a multiplication of opportunities to initiate independent clones. But what about stalks and pigment? In the fungi, the spores can become airborne and thus represent a dispersal mechanism. It seems obvious that if the spores are elevated from the substrate, the opportunity for them to be wafted off is increased. In the case of the myxobacteria, there is no evidence that this is the case. The spores or the cysts are usually embedded in a slime matrix and seem not to possess the highly hydrophobic character of spores that spend a great deal of their time exposed to the elements. It seems as if the explanation will turn out to be more subtle, and no insights as to its nature are presently available.

H. Conclusions

One of the intellectual advantages of studying bacterial development is that most of the developmental changes that bacteria undergo can be rationalized in terms of increased adaptability—for example, changed resistance, metabolic diversity, or enhanced food-gathering. In only one of these categories—sporulation—are there models that attempt to explain the regulatory couple between the environmental condition and the developmental event. And these models, which are based on nucleotides as messengers between the environment and developmental biosynthesis, concern only one corner of the process. The question of how the environmental circumstances are perceived and translated into the synthesis of the nucleotide messengers is unclear, as is the transduction of the nucleotide signals into actual developmental events. With regard to the other categories of developmental events, they lie fallow.

Chapter 9

The Molecular Basis of Morphogenesis

The organized growth of an embryo from a single fertilized egg to a functioning adult has fascinated biologists from ancient times to the present. Although the successive events of embryonic development in many species have been described in precise detail, the fundamental question of embryology still challenges the most sophisticated resources of modern biology: What controls the process of morphogenesis, or the origin of form? In other words, how does a one-dimensional genetic code ultimately specify a three-dimensional organism?[1]

A.   Introduction

The theme of this chapter comprises the following questions:

1. How can the shape of a bacterial cell or of one of its organelles be explained in terms of its chemical composition? Thus, for example, how can one describe the helical nature of a flagellum in terms of the structure of its subunit protein, flagellin? Or how can one explain the spherical shape of *Staphylococcus* in terms of its cell wall structure? Or how can one explain the rounded contour of the end of a rod versus the linear quality of the rest of the cell surface? These attempts to describe the simple structural features of a nondeveloping cell are, of course, only a prelude to more ambitious efforts to

understand the structures of spores, cysts, stalks, fruiting bodies, and so on.

2. As cells *change* shape during development, how can these shape changes be described in chemical terms? Here the key word is *change*, for it adds the dimension of time to question number one.

3. How are these changes regulated in the context of the developmental cycle?

In 1917, the great British naturalist D'Arcy Thompson published his classic work *On Growth and Form*.[2] In it, he attempted to remove biological shape and form from subjective and imprecise analysis and instead to subject it to mathematical and physical description and analysis. Thompson was not concerned with causal explanations; he was satisfied merely to describe in an analytical fashion. Now, almost 70 years later, we are able to describe many biological objects in chemical fashion, but the causal connection between the chemistry and the form continues to elude us, or to paraphrase Whitehead, "the darkness of the subject remains unobscured." Cell shape determination is an area of development that receives scant attention; in contrast to the hundreds of laboratories concerned with the question of differential gene expression during development, one can count the laboratories actually working on developmental morphogenesis on one's fingers. A major reason for this small number is probably that the conceptual underpinning for relating chemical structure to actual form is a fledgling science. Nevertheless, the areas in which one is able to study the problem and, in some cases, actually draw some insight are:

a. the role of peptidoglycan structure
b. flagellar self-assembly
c. ribosomal self-assembly
d. polysaccharide self-assembly

B. Peptidoglycan and Cell Shape

In bacteria, the molecule that offers the greatest opportunity for relating structure to shape is, of course, the peptidoglycan. The structure for the peptidoglycan of *Escherichia coli* is illustrated in Figure 9-1. There is a great deal of evidence that indicates that the shape of bacterial cell could be attributed to its peptidoglycan. For example, if one disrupts the peptidoglycan of a rod-shaped cell either with antibiotics or enzymes, the cell, released from its shape-determining corset, assumes a spherical shape. Even

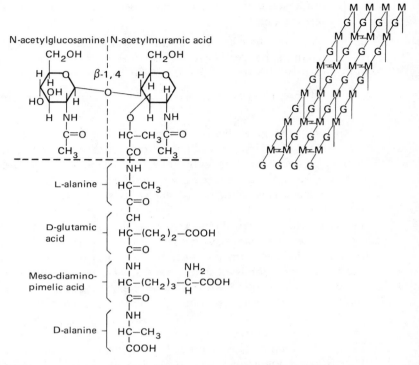

Figure 9-1 Diagram illustrating the chemical structure and physical organization of cell wall peptidoglycan of *Escherichia coli*. G and M designate residues of N-acetylglucosamine and N-acetylmuramic acid, respectively. The lines extending from M represent tetrapeptides attached to muramic acid residues. (Adapted fdrom Ghuysen 1968)

more convincing is the fact that it is possible to isolate pure peptidoglycan from rod-shaped cells that retains the rod-shaped configuration of the cell itself (see Figure 4-5). In fact, peptidoglycan has been referred to as a bag-shaped macromolecule. Thus, the simplest view of shape determination in bacteria is that it is inherent in the molecular structure of the peptidoglycan itself; and variations in the molecular nature of the subunits or in their assembly patterns can determine the three-dimensional configuration of the cell.

There have been numerous attempts to relate cell shape to particular aspects of peptidoglycan structure. One of these attempts suggests that the length of the polysaccharide backbone can determine the rigidity of the structure and thus the shape of the cell. Unfortunately, analyses of the peptidoglycan of bacteria with different cell shapes have not supported this notion. Another suggestion has been that the degree of cross-linking between the peptide chains is a shape-determining factor. While, in its simplest form, this suggestion has not been borne out, E. P. Previc, in a very thoughtful review,[3] has suggested a more sophisticated version of this idea. Previc claims that existing data indicate that a correlation exists between the shape of bacterial cells and whether their peptide contains lysine or diaminopimelic acid (DAP); according to Previc, lysine occurs in the peptidoglycan of most spherical cells whereas DAP is found in rod-shaped cells. Lysine is trifunctional (two amino groups and one carboxyl group). DAP is tetrafunctional (two amino groups and two carboxyl groups).

Previc argues that the cylindrical portion of the rod-shaped cells needs to be more rigid than the surface of a sphere. In the latter case, if the normal osmotic forces within a cell are evenly expressed over the enclosed plastic surface, that surface will assume a spherical shape. Thus, a deformable but nonexpandable surface is required. On the other hand, the surface of the cylinder must be nondeformable except at the rounded ends. Previc suggests that trifunctional lysine allows establishment of the deformable sphere surface and that tetrafunctional DAP allows sufficiently additional cross-linking to establish the more rigid complex, nondeformable rod, with only some of the functional groups used to establish the surface of the cell ends. It is an interesting idea but the data do not support it in that only three of the DAP functional groups are found to be involved in covalent bonding. Previc suggests that this is because cells that are thus analyzed are usually rapidly growing cells whose peptidoglycan is continually being opened and closed so that inserted fragments can allow for cell surface growth. Another serious objection is that the correlation is not at all as close as suggested; essentially, all Gram-negative bacteria contain DAP; among the Gram-positive bacteria, some species of *Lactobacillus, Corynebacterium,* and *Bacillus* contain lysine while some species of *Micrococcus* contain DAP.

Daneo-Moore and Shockman have for many years studied the biochemistry of peptidoglycan and the cell surface with regard to the processes of growth and cell division and are convinced that one must seek more subtle explanations for shape control than peptidoglycan structure.[4]

It has become clear that this simple view of shape determination is inadequate. What then is the role of the peptidoglycan in the shape of the cell? It is necessary to distinguish two aspects of the problem: the determination of shape and the maintenance of shape. They are not the same and in bacteria are most probably separate processes. Imagine that it is necessary to mold a malleable, deformable object into a particular shape. Let us say one wished to convert a round balloon into a rod-shaped one. One could enclose the round balloon in a cylindrical mold and then pour plastic around the balloon. Upon removal from the mold, it would be clear that the plastic was responsible for maintaining in a rod shape an object that, if freed of external constraint, would be round. The plastic has not determined the rod shape, but it is certainly maintaining it. Likewise, if we think of the plastic as the cell wall peptidoglycan and the balloon as the cell protoplast, the analogy allows us to intuit the maintenance function of the peptidoglycan, but it certainly does not lead us to think of any shape-determining function.

With this distinction in mind, one can look considerably more usefully at the relation between peptidoglycan structure and its function. The peptidoglycan layer must be semirigid and expandable at any point of its surface by the insertion of new subunits. One would not predict that there need be any molecular distinction either between the subunits of round or rod-shaped peptidoglycan or necessarily between the hemispherical ends versus the cylindrical sides of a rod. It is possible that the molecular variations among peptidoglycans are more related to the need for regulatory road signs for the hydrolytic or biosynthetic peptidoglycan enzymes than for some biomechanical property of the molecule.

What then are the forces that determine cell shape? And here, of course, the mystery deepens, and one may suggest, rather lamely, that the topographical organization of the biosynthetic process itself—the three-dimensional disposition of the peptidoglycan synthesizing enzymes themselves—represent the blueprint that determines the shape of the cell. Or perhaps a three-dimensional template would be a more useful metaphor.

In a provocative series of papers,[5] Arthur Koch has taken a more specific and biophysical approach and has proposed that the size and shape of bacterial cells is determined by hydrostatic forces inside the cell and surface tension forces on the outer surface of the cell. The problem faced by a cell taut with hydrostatic pressure but needing to expand its surface to accommodate a growing cytoplasm is essentially the same as that faced by engineers attempting to dig a tunnel through mud, build a dam, or patch a hole in a ship's hull. The solution proposed by Koch is a simple one; the cell attaches a new segment of peptidoglycan to the old surface before severing the bonds that bear the stress in the old sacculus. When the new unit of peptidoglycan has been attached to the old by at least two covalent bonds, the appropriate peptide bonds in the old segment are cleaved; the ensuing expansion on the old, external wall, caused by the release of tension pulls the newly inserted unit into the wall. This process is illustrated in Figure 9-2.

The shape of the cell is determined by where the new growth takes place. For a Gram-positive coccus such as *Streptococcus*, Koch proposes that the growth is zonal and confined to a narrow region near the developing septum. For a Gram-positive rod, the appropriate shape can be generated and maintained if there are three types of growth zones: (1) on the inner edge of the ingrowing system, (2) at the junction of the septum and the newly forming pole, and (3) along the cylindrical walls. For the last, the model requires that the growth be diffuse throughout the wall. A gram-negative rod such as *E. coli* requires a special model, as its growth differs from that of *Bacillus subtilis*. The cell is not, strictly speaking uniformly rod-shaped, and its shape and maximum diameter vary during the growth

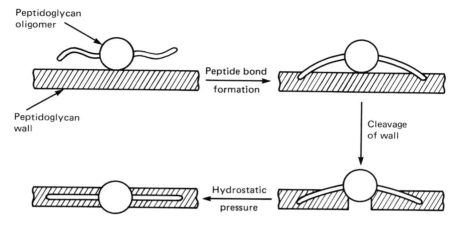

Figure 9-2 A model suggesting how a structure under tension (e.g., the cell wall) can be opened during growth to allow the addition of a new piece without bursting the cell. (Adapted from Koch et al. 1981)

cycle and with the nutritional milieu. This lack of uniformity requires a somewhat more complex model where the surface tension of the cell surface changes during the growth cycle.

In other words, Koch proposes that the primary forces that determine the shapes and sizes of bacterial cells are not necessarily enzymatic or regulatory but rather stem from hydrostatic and surface tension effects. Thus, there is no need to involve either the elaborate scaffolding mechanisms shown to be involved in the formation of virus heads nor the inherent molecular properties of the structural building blocks. While these most certainly play a role in the process, it is Koch's important point that the primary forces are physical ones.

This entire notion of construction oriented by some sort of three-dimensional template stands in contrast to the processes of shape determination manifested by viruses, flagella, or ribosomes. In these cases, the determination and maintenance of shape are controlled precisely and strictly by the molecular conformation of the subunits. Change the primary structure of the protein of a virus head and the self-assembly process is thrown awry. Either it fails to assemble or it assembles as a misshappen mutant.

C. Flagellar Self-Assembly

Flagella from both Gram-positive (e.g., *Bacillus subtilis*) and Gram-negative (e.g., *Salmonella*) bacteria can be dissociated to their flagellin protein subunits. These subunits can then be reassociated in vitro to form intact flagella. Thus, it is possible to examine the relationship between flagellin composition and structure and flagellar morphology. For example, substituting amino acid analogues for the normal amino acids has been shown to generate changes in the flagellar filament.[6] Furthermore, it has been possible to explain the superhelical form that flagella may assume in terms of orientation of flagellin subunits within the filament.[7]

D. Ribosome Assembly

Ribosome assembly also involves the self-assembling properties of proteins. There are approximately 20 operons coding for 53 ribosomal proteins. A set of proteins binds to the 16S RNA molecule in a self-assembling process that can take place at 0 C. This complex then must undergo a temperature-sensitive change (perhaps enzymatic) after which it can then self-assemble an additional group of proteins. The final, complete 30S ribosomal subunit then emerges. The assembly of the 50S particle is less well defined but follows essentially the same process involving self-assembly onto a 28S and 5S RNA complex, temperature-sensitive processing, followed by a final, additional self-assembly resulting in the 50S subunit.[8]

Up to this point, the chapter has focused on those shape-determining systems that have been subjected to some kind of analysis—theoretical or experimental; the rest of the chapter shall focus simply on describing the chemical composition of the various cell surface structures that seem to be involved in morphogenetic changes.

E. Bacillus Endospore

There are three surface layers of the endospore that could play a role in spore morphogenesis: the spote coat, the cortex, and the spore membrane. Each of these has been described in Chapter 3.

Based on the exceptional chemical resistance of isolated spore coats, it is tempting to conclude that the spore coat plays some role in spore resistance. There is now a fair amount of evidence, using mutants deficient in coat synthesis, that this is not the case. Such mutants retain normal resistance to heat and ultraviolet light irradiation. It is, however, interesting that they become sensitive to lysozyme digestion and become deficient in their ability to respond normally to germinants. Thus, the spore coat seems to represent a physical barrier to access to the underlying peptidoglycan as well as a site for germination receptors.

As pointed out earlier, the cortical peptidoglycan differs substantially from that found in the vegetative cell wall or for that matter from peptidoglycan in general (Figure 9-1). Two of the possible functions that can be attributed to the cortical peptidoglycan are a role in the establishment and/or maintenance of the state of partial dehydration of the spore protoplast. With regard to shape determination, isolated cortical peptidoglycan does in fact, assume the spherical shape characteristic of the spore. However, the question as to which is the cause and which is the effect of the spore shape cannot at this point be resolved. Two differences between spore and vegetative cell peptidoglycan are worth considering. In *B. subtilis* vegetative cells, 60% of the peptidoglycan is cross-linked; in the spore cortex, only 6% of the disaccharides are cross-linked. This would permit the peptidoglycan to assume a deformable spherical shape in contrast to the more rigid cylindrical shape characteristic of the vegetative cell. In *Bacillus sphericus*, the vegetative peptidoglycan is cross-linked between lysine and D-alanine via D-isoasparagine residues; in the spore cortex, the cross-links are between DAP and D-alanine; lysine and isoasparagine are absent.

The role of the cortical peptidoglycan in spore resistance via protoplast dehydration has been a debated issue for many years since it was originally proposed by Lewis, Snell, and Burr in 1960.[9] These authors proposed that the cortex contracted around the spore protoplast, in effect squeezing some of the water out of it. Subsequent theories have proposed that the cortex would act as an osmoregulatory organelle. The presence of a high concentration of cations in the cortex would generate an osmotic pressure and result in the flow of water out of the spore protoplast and into the cortex. This theory is interesting and ingenious—despite the fact that much current opinion seems to favor the idea that resistance and dormancy are based on noncovalent interactions between spore macromolecules and calcium dipicolinic acid. These interactions are believed to result in a solid support system that reduces the molecular mobility of spore constituents, thereby increasing their stability and resistance. The role of the cortical peptidoglycan as the physical matrix for this system represents a fascinating problem in morphogenesis.

F. Cyanobacterial Heterocysts

Chapter 5 has already discussed the nature of the cyanobacterial heterocyst and its function as a site for dinitrogen fixation. From the point of view of morphogenesis how can one relate the cell's structure to its function? The heterocyst must have the following properties:

1. It must be able to fix dinitrogen.

2. It must be able to protect the nitrogenase complex from ambient or metabolic oxygen.

3. It must be able to export fixed nitrogen and import fixed carbon.

4. It must be properly spaced along the filament.

From a structural point of view, there is little to be said about point 1. There seems not to be any discrete structures that contain or localize the nitrogenase complex. The enzymes are made in the heterocyst percursor, the proheterocyst. The salient developmental question in this regard is, what is the nature of the regulatory mechanism that couples proheterocyst formation with the expression of the nitrogenase genes?

The matter of oxygen protection by the heterocyst is an interesting one. There are apparently a number of features to the protection, some of which are unrelated to heterocyst structure. For example, the nitrogenase enzyme appears in the proheterocyst before the characteristic heterocyst morphogenesis has occurred; presumably the nitrogenase complex is stabilized in some fashion not involving a physical barrier to oxygen. While there is no direct evidence to bear on the question, Peter Wolk has suggested that the glycolipid layer that comprises the innermost of the three heterocyst layers can sufficiently retard the diffusion of oxygen into the cell so that resident scavenging enzymes can dispose of the oxygen that does find its way inside.[10]

With regard to the import/export problem, it has been suggested that either there are specific pumps at the junctions of vegetative cells and heterocysts that control this interaction or that the microplasmodesmata (channels traversing the septal junctions) may selectively allow the one-way passage of nitrogen or carbon. There is essentially no information that bears directly on this aspect of heterocyst morphogenesis. Finally, as pointed out in Chapter 5, the issue of heterocyst spacing remains a fascinating and unsolved problem. However, the process concerns morphogenesis per se less than it does the role of diffusible signals, cell interactions, and turning processes on and off. These points will be discussed in subsequent chapters.

G.   Myxobacterial Spores and Fruiting Bodies

There are four aspects of myxobacterial morphogenesis where a start has been made and where continued analysis might lead to some insights into the problem:

1. Examination of vegetative cell and spore peptidoglycan.

2. Characterization of the myxospore coat.

3. Protein S on the surface of fruiting body myxospores.
4. Analysis of the polysaccharides of fruiting bodies.

1. Peptidoglycan

The peptidoglycan of *Myxococcus xanthus* is especially interesting because it seems to exist not as a continuous layer but rather as patches connected by a material sensitive to trypsin and detergent.[11] There are two aspects of the life style of this myxobacterium to which the fact of a patchy peptidoglycan may be relevant. The vegetative cells are rather flexible; under some conditions they can bend into hairpin loops. This flexibility could be either a trivial reflection of the fact that the cell is rather long and thin, or it could be a functional adaptation to the characteristic gliding motility. It may be necessary for the cells to fit the contours of the surface on which they are gliding. Alternatively, the patchy peptidoglycan might facilitate the shape change of the vegetative cell to the round myxospore. The peptidoglycan of the resistant myxospore differs significantly from that of the vegetative cell; it is more highly cross-linked, and it is no longer sensitive to disaggregation by trypsin and SDS. It is also interesting that during the period of cellular morphogenesis when the rod-shaped vegetative cell begins to round up, there is a temporary decrease in the degree of cross-linking in the peptidoglycan. One must resist the temptation to draw the easy conclusion that shape determination and perhaps also the resistance of the myxospore is related to the peptidoglycan cross-linking. At the moment, let's just say that it may be part of the story.

2. Myxospore Coat

As indicated in Chapter 7, the principal component of the myxospore coat is a protein-polysaccharide complex that may be covalently associated as a glycoprotein.[12] Chemically, the coat is extraordinarily tough and may play a protective role in the myxospore. If the coat polymer is indeed a glycoprotein, it raises the interesting morphogenetic question as to how the protein-polysaccharide bonds are established outside the cell and separated from the cell membrane by the peptidoglycan layer. David White has made the interesting suggestion that carbohydrate and protein portions of the spore coat may derive from the interpatch material just referred to. He has proposed that spore morphogenesis may involve the evagination of the interpatch material, bringing the peptidoglycan patches closer together. This action would be followed by increased peptidoglycan cross-linking and finally by the breaking of the bonds in the interpatch material, allowing it to become part of the coat.[13]

3. Protein S

Myxospores of *M. xanthus* synthesize a protein termed "Protein S" that seems to self-assemble into a 30-nm layer external to the spore coat and only in fruiting-body myxospores. It is not present on glycerol-induced myxospores nor will it self-assemble on them. Protein S is made late in the process of spore formation and does not seem to be involved in the acquisition of resistance to temperature extremes or to sonic oscillation. There is a suggestion, however, that it serves as an intercellular adhesive to keep the spore tightly packed in the fruiting body. The molecule has been thoroughly characterized and offers the opportunity to examine the regulation of a defined, specific, developmental gene product.[14]

4. Fruiting Body

Who can have examined the elaborate regular fruiting bodies of the myxobacteria and not have wondered how a simple bacterial cell could have constructed such a splendidly complex structure? As indicated earlier (Section F of Chapter 7), I think that part of understanding the phenomenon lies in the splendid work of David Rees on polysaccharide self-assembly.[15] Polysaccharide chains are able to enter into reversible interactions with each other so that it is possible to alternate between sol and gel states. The interactions between chains results in a cross-linked, three-dimensional network with the properties of a gel. Each chain is a heteropolymer where stretches of one subunit are interrupted by stretches of another. This discontinuity allows portions of one chain to interact with portions of another, leaving the remainder of the chain to interact with yet other chains. These interactions can take a number of different forms, illustrated in Figure 9–3, and in some cases, alginates, for example, the gel formation is catalyzed by divalent cations such as Ca^{++}. The gel can be melted by an ion-exchange reaction where the Ca^{++} is replaced by Na^+. Thus, the possibility exists that a cell could produce the proper mixture of polysaccharides that remains fluid until the proper moment, at which time the appropriate divalent cation is made available and the mixture gels—that is, it forms "structure." The ability of myxobacteria to construct their elaborate fruiting bodies may therefore reflect their ability to order the synthesis and excretion of the proper mix of polysaccharides plus the appropriately timed efflux of divalent cations to catalyze gelation.

H. Lessons to Be Learned

1. The shape of various bacterial organelles has to be examined both in terms of determination and maintenance.

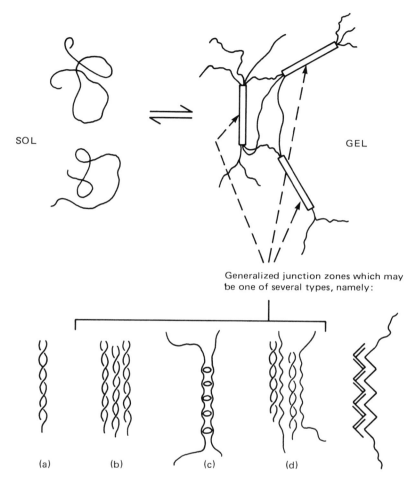

Figure 9-3 This illustration shows the kind of intermolecular interactions in polysaccharides that can lead to reversible gel-sol transformations: (a) cross-linkages by double helices as in carageenan; (b) by bundles of double helices as in agarose; (c) by ribbon-ribbon associations of the egg-box type as in alginate; (d) helix-ribbon associations as in mixed systems containing agarose or certain mannans or xanthan (a bacterial polysaccharide) and certain glucosamines. (Adapted from Rees 1977)

2. The only cases where one understands the basis for shape determination are those involving self-assembly of virus particles, flagella, or ribosomes. In those cases, the shape of the organelle and its biomechanical properties are an explicit function of the nature and packing of the polymer subunits.

3. In the case of cellular shape, its maintenance is most likely a function of the nature of the peptidoglycan, although probably relatively

independent of such variations as acetylation, cross-linking, length of cross-bridges, length of glycan backbones, and so on. The function of such variations is not understood. The determination of cellular shape is probably a more subtle reflection of the three-dimensional architecture of the biosynthetic enzymes involved in laying down the peptidoglycan polymer.

4. The self-assembling properties of certain mixtures of polysaccharide offers the interesting opportunity to think about shape determination by polysaccharides. In fact, peptidoglycan has been referred to as a bag-shaped macromolecule. Thus, the simplest view of shape determination in bacteria is that it is inherent in the molecular structure of the peptidoglycan itself; and variations in the molecular nature of the subunits or in their assembly pattern can determine the three-dimensional configuration of the cell.

Chapter 10

Regulation of Development

A. Introduction

Development, like all other biological processes, is the result of a sequence of metabolic events. A central question, therefore, concerns how those metabolic events are organized and regulated individually and collectively.

In a sense, modern regulatory biology began with Beadle and Tatum's work on the regulation of biosynthesis in *Neurospora*. Since then, the models that have been most productively examined have been such systems as amino acid biosynthesis and the lac operon. In such cases, regulation is a spatially linear, one-dimensional phenomenon. Positive and negative feedback loops, amplifications, and modulations all collaborate to determine how much of a particular end product is synthesized or which alternative pathway is emphasized. The nature of the process being regulated is such that relatively unstable or (perhaps more accurately) more easily reversible control mechanisms are called for. The cell must respond to changes in its environment that are transient. In order to balance continuously one biosynthetic sequence with another or to respond to the availability of energy, the cell must be constantly fine tuning. Development, on the other hand, must involve a set of regulatory processes that have a more metastable quality about them—and, in fact, such processes as RNA polymerase modifications, gene inversions, and transpositions are beginning to emerge as focal topics for developmental regulation.

B. The Timing of Developmental Events

An underlying feature that distinguishes development from many other biological processes is the dimension of time. The issue of the nature of temporal regulation in development is an interesting question that has not yet really been resolved. If one begins with the premise that all developmental change stems from the differential expression of developmental genes (this position, while held by the overwhelming majority of workers in the field is not held universally; see, for example, Barbara Wright's thoughtful opposition to this view[1]), then the question is, how is the sequence of regulatory events controlled? There are two general alternatives that have been described as the metabolic model and the genetic programming model.

In the metabolic model, sequential changes in the internal environment of the cell are generated by metabolic reactions. These changes act as the triggers for the appropriate gene activation. The timing then is a result of the time required to generate and accumulate the necessary triggering molecules. In the genetic programming model, the assumption is that the instructions for timing the expression of the various developmental genes are themselves encoded in the DNA. Which of these two models is correct? Are they both involved in temporal regulation? By what mechanisms would temporal genes operate? These are fundamental and provocative questions; while they remain essentially unanswered there are some clues that are available based on the study of certain mutants of mammalian cells.[2]

These mutants are temporally deranged and are particularly useful because the mutants are defined not by some vague developmental phenotype but rather by the timing of the appearance or disappearance of specific enzymatic or antigenic proteins. These include β-galactosidase, glucuronidase, H2 histocompatibility antigens, and a number of other proteins. Genetic analyses of these mutants suggest that temporal expression of these mutant properties is a programmed process rather than a consequence of a series of related metabolic events. Thus, while there is indeed some evidence for programmed, temporal regulation, it is difficult to imagine that this occurs in a fashion insulated from the internal or external environment of the cell. There must indeed be an interplay between the two modes, but, alas, the extent and nature of the interplay are complete mysteries.

There has been relatively little attention paid to temporal regulation of prokaryotic development. It has been shown, however, that in *Myxococcus xanthus*, cell division requires that a small piece of the chromosome (about 5%) near the initiation be expressed. In addition, the cell must complete DNA replication. Furthermore, during myxospore formation, the protein gene product of this early gene expression is apparently destroyed, so that subsequent germination requires not only that replication near the initiation take place but also that the putative protein gene product be resynthesized.[3]

The link between DNA synthesis and the timing of developmental events has been studied in *Caulobacter,* and a model has been proposed in which DNA replication is considered as a cellular clock that controls the division pathway, periodic flagellin synthesis, hook protein synthesis, and the expression of phage receptor sites in stalked cells.[4] The conclusions are based on an ingenious technique for ordering the steps in a process by means of hot and cold temperature blocks in hot and cold temperature-sensitive mutants.[5] The idea that DNA replication acts as a timer for cellular events is not a new one; it has been proposed for yeast as well as for other cells.[6] It may turn out to have considerable general validity when considering the timing of developmental events.

C. Spatial Aspects of Development

Another aspect of development that adds to its distinctiveness and to its complexity is the spatial dimension. Many years ago, Peter Mitchell put forward the idea that metabolism had a vectorial as well as a scalar dimension.[7] In other words, metabolism has physical direction. This notion has been particularly useful when considering events that occur in or through the cell membrane, as it is possible to visualize the membrane as a matrix in which enzymes are embedded and have a functionally specific orientation.

As indicated in Chapter 4, development in *Caulobacter* involves the spatial and temporal ordering of surface structures—for example, stalk, flagellum, pili, and phage receptor sites. While there is not much information that bears directly on the question of how this spatial orientation is organized, Austin Newton's laboratory at Princeton has recently proposed a somewhat more explicit model. Their work had previously demonstrated that (1) assembly of new polar structures is always initiated at the newest cell pole (Figure 10-1); (2) formation of these polar sites requires that the cell have completed a late step in its previous cell division pathway; and (3) the old assembly site—the previous organizational center—is inactivated during the swarmer-stalked cell conversion. They thus propose that there exist "organizational centers" that are laid down during the previous cell cycle.[4]

With regard to the role of these centers, they have recently obtained evidence that indicates that the polar portion of the cell membrane of *Caulobacter* differs in its protein composition from the rest of the cell membrane. It appears that these polar segments function as the site of insertion of the flagellin molecules into some kind of a membrane pool. Lucy Shapiro has suggested that the temporal and spatial aspects of development

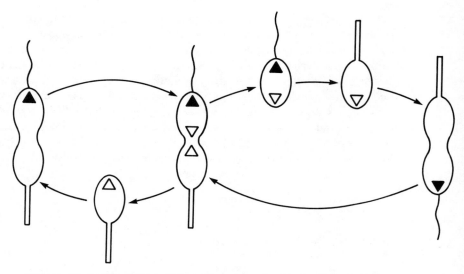

Figure 10-1 Model describing the proposed role of cell division steps in the assembly of polar structures in *Caulobacter*. The hypothetical organization centers are laid down late in cell division at the new cell poles (△). In the next cell cycle these centers become the active sites (▲) for assembly of the polar flagellum and phage receptors. The assembly sites are then inactivated later during the swarmer cell–stalked cell conversion. (Adapted from Huguenel and Newton 1982)

in *Caulobacter* may be jointly controlled.[8] The observation has been made that the physical properties (e.g., the sedimentation coefficient) of the membrane-associated nucleoid changes during the cell cycle. Shapiro suggests that the site on the membrane at which the nucleoid is located changes during the cell cycle and that these changes are functional in that they determine the membrane location at which a specific organelle will be synthesized. The mechanism whereby this occurs could be a posttranscriptional one where the ribosome-mRNA complex coding for a particular organelle component can only form or perhaps bind to the membrane at a particular membrane site. Since it has already been shown that the temporal expression is controlled in some fashion coordinate with DNA replication, this model has the virture of combining both the temporal and the spatial regulation in one process. Of course, what controls and activates the changing location of the nucleoid on the membrane is unknown.

D. Transcriptional Regulation of Development

In the area of the mechanism of regulation of gene transcription during bacterial development, the ratio of facts to speculations begins to climb.

And here, our insights are largely attributable to the efforts of Richard Losick at Harvard. Some of this work has already been briefly referred to in Chapter 3 during the discussion of the control of transcription during *Bacillus* sporulation.

In *Bacillus*, a major site of regulation of the expression of developmental genes seems to be the accessory sigma factors that are required for the operation of RNA polymerase.[9] These sigma factors are proteins of varying sizes that apparently recognize characteristic, brief sequences of nucleotides on the promoter upstream from the site of actual transcription. Whether or not these sites are recognized by sigma factors determines whether or not the genes regulated by the promoter will be transcribed by the RNA polymerase. In *Bacillus*, there are at least five species of sigma factors known as σ^{55}, σ^{37}, σ^{32}, σ^{29}, and σ^{28}. The specificity of each was determined by in vitro tests with cloned gene fragments, such as cloned sporulation genes. The most extensively studied case is the sporulation gene known as *spoVG*, a mutation which causes a block at a late stage of sporulation even though its transcription commences early after the onset of development.

Interestingly, *spoVG* is expressed from two overlapping promoters that are separately recognized by σ^{37}- and σ^{32}-containing forms of RNA polymerase. In order to rationalize the involvement of these RNA polymerase forms in sporulation, Losick proposes the following model. The products of *spo0* genes, a class of regulatory genes that are responsible for triggering the onset of sporulation, are activated by some signal associated with nutritional deprivation. The *spo0* gene products then interact with σ^{37} and σ^{32} RNA polymerases and thereby stimulate transcription from early expressed sporulation genes, such as *spoVG*. The gene products resulting from these events interfere with the binding of the vegetative sigmas, σ^{37} and σ^{55}, to the polymerase. This phenomenon may be the inhibition of the σ-polymerase binding referred to earlier in Chapter 3. Thus, vegetative and early sporulation genes are turned off as sporulation progresses. Finally, σ^{29} appears and replaces σ^{37} and σ^{55}, resulting in the continuing transcription of some vegetative genes and early sporulation genes as well as turning on the stages II and III sporulation genes.* Figure 10-2 is a polyacrylamide gel electrophoresis showing three sigma factors of *Bacillus subtilis*. There are obviously many gaps in the model, but it provides a conceptual framework that rationalizes the regulatory and temporal aspects of a developmental process. It is a lovely piece of work.

Polymerase modifications create an additional regulatory problem for the cell. If the specificity of the polymerase is modified by a different accessory factor, how does the cell continue to transcribe those genes whose expression is needed both for vegetative growth and during development? A strategy is beginning to emerge that is exemplified by both *Bacillus* and

---

*Recently it has been shown that σ^{29} appears only briefly (2-4 hr after the initiation of sporulation);[10] its appearance seems to set off an entire cascade of developmental events.

Figure 10-2 SDS polyacrylamide gel electrophoresis showing the three sigma factors of Bacillus subtilis, σ^{29}, σ^{37}, σ^{55}. (From Losick 1981)

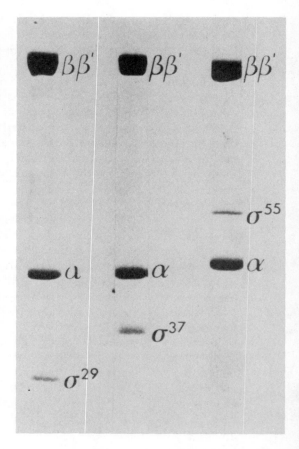

Anabaena. The solution for the cell seems to be to allow the transcription of such genes to be controlled by multiple promoters. Thus, the same gene may be controlled by two promoters that are either overlapping or separated, each of which has a specificity for a different accessory factor. In the case of Bacillus, transcription of the spoVG gene is controlled by two overlapping promoters, each of which responds to a different sigma factor. In the case of Anabaena, the cell apparently has to synthesize glutamine synthetase whether it is using a fixed nitrogen source or is fixing molecular nitrogen. However, which of those two circumstances exists has a determining effect on the cell's developmental stage—in the absence of fixed nitrogen, the organism begins to undergo heterocyst development. Thus, certain cells must turn on the genes for heterocyst development and nitrogen fixation and must also turn off those genes involved in CO_2 fixation and oxygen evolution. The cell must nevertheless continue to synthesize glutamine synthetase. Haselkorn's group has cloned the glutamine synthetase gene, and using it as a probe for the corresponding mRNA has shown that the same glutamine synthetase gene is transcribed from different promoters during growth on fixed nitrogen or molecular nitrogen. In the former case, the promoter sequence resembles the consensus sequence of *Escherichia coli* promoters; in the latter case it resembles the promoter sequence of the nitrogen-fixing genes of *Anabaena*.[11]

Thus, the existence of multiple promoters with different recognition specificities controlling expression of the same gene allows the cell to continue to transcribe required genes even though the specificity of the polymerase has shifted from a growth to a developmental mode.

Neither the polymerase itself nor the accessory sigma factors in *Caulobacter* were found to change during the cell cycle;[12] thus, differential gene expression during *Caulobacter* development seems not to be a function of any changes in RNA polymerase. The RNA polymerase of *M. xanthus* has been characterized and seems to contain one or more sigma-like factors;[13] it has not yet been possible, however, to examine the polymerase of developing cells. However, in *Streptomyces coelicolor* the RNA polymerase has recently been shown to be heterogeneous, and there are two species of sigma factor that recognize different promoter classes.[14] Whether or not this fact is developmentally significant in *Streptomyces* is not yet evident; if it is, then the control of differential gene expression by sigma factor promoter recognition may represent a general developmental paradigm.

E. Translational Control of Development

There is considerably less experimental evidence bearing on posttranscriptional levels of developmental regulation. However, it is clear that such control does exist.[15] This control could be a result of stable mRNA's, ribosome modifications, altered tRNA's, or all of the above. One strategy for demonstrating that translational-level control exists has been to use inhibitors that interfere with ribosomal function. For example, several hundred single-site mutants of *B. subtilis*, resistant to the antibiotic erythromycin, have been isolated.[16] This antibiotic interferes with protein synthesis by binding to the ribosomes and preventing translocation; thus, incomplete peptides are formed and the polysomal complex is destabilized. Resistance to the antibiotic is a result either of a modification of the L17 protein of the 50S subunit of the ribosome or of methylation of the 23S RNA. The interesting feature of these resistant mutants of *B. subtilis* is that all of them were also temperature sensitive for sporulation—that is, they were unable to sporulate at slightly elevated temperatures. This feature suggested that the modification that resulted in erythromycin resistance (in this case it was modification of the ribosomal protein) also interfered with sporulation. In other words, alterations of the ribosome that have no effect on vegetative growth prevented sporulation; either the ribosomes that are involved in synthesis of sporulation proteins are different from those in the vegetative cell or different portions of the ribosomal particle are involved in the two processes. It should be kept in mind, however, with all these sorts of experiments that an alternative explanation is possible—the mutant ribosomes translate less efficiently than wild-type ribosomes. If the requirements for protein synthesis during sporulation are more demanding than during growth, this aspect would be reflected as a specific effect on sporulation.

If an organism can synthesize a stable mRNA (one that does not break down with the usual half-life of a few minutes), but instead persists for a long period of time, then the level of control is shifted from the transcriptional one to the translational. While the battlefield is littered with the corpses of investigators claiming to have demonstrated a stable mRNA in bacteria, it is now beginning to appear as if some bona fide cases do indeed exist. Recently, it has been shown that *M. xanthus* during its early stages of development synthesizes stable mRNA and that this mRNA codes for the S protein, synthesized later in development.[17] Presumably, the stable mRNA can be stored in the cell in preparation for subsequent use. The questions are, of course, what is the mechanism whereby the normal turnover is arrested or retarded? what controls whether or when the mRNA will be translated? The notion of stable, stored mRNA is not new; various eukaryotes such as sea urchin eggs, silk worms, and various fungi have long been known to make use of stable mRNA, and mRNA's with somewhat extended half-lives have been reported for a few prokaryotes.

Joel Mandelstam's laboratory at Oxford has long been interested in the temporal relationship between the transcription of a developmental gene and its final, phenotypic expression. His laboratory has examined a number of sporulation and germination properties of *B. subtilis*.[15] They have found that the development of such properties as heat resistance and resistance to toluene could not be prevented by the addition of inhibitors of protein synthesis, even though those inhibitors were added before the actual phenotypic appearance of the property. The same strategy was used to demonstrate that the actual germination properties of the spore also emerged some time after the proteins necessary for these properties had been synthesized. In other words, there is a temporal gap between the expression of the genes and the actual emergence of the phenotypic property. Any substantial delay between the synthesis of a specific messenger and the appearance of the corresponding protein requires that there has been some regulation exerted at the translational level of control. It has been a matter of implicit dogma that the sequence of developmental events reflects the sequence in which the corresponding developmental genes are expressed. Mandelstam's work shows that this is an oversimplification, that processes such as posttranslational processing or self-assembly into higher order structures may play a role in the process and that these events themselves must also be subject to some developmental program.

There is ample evidence that the sequence of many sporulation events does in fact reflect the ordered, linear expression of a set of equivalent genes; there is also evidence that substantial parts of development do not — that the regulatory network is a mosaic of criss-crossed, multidimensional interactions. A highly simplified but extremely useful model to illustrate both the linear-dependent quality of the process and interpathway connections has been proposed by Piggot et al.[18] (Figure 10-3):

> In this model, the switching on of successive operons via elements C_a, C_b, C_c, etc., could be mediated by low-molecular-weight products of enzyme action (as S_3 on C_b and C_c) or by protein (p_3 on C_c). Different consequences of point mutations that inactivate the corresponding gene product might be anticipated. In operon A, mutation in G_1 or G_2 would prevent expression of operon B and consequently subsequent operons; the resulting phenotypes would be difficult to distinguish unless, by chance, any of P_1, P_2, S_1, S_2, or S_3 was known. In operon B, mutation in G_3 would prevent expression of subsequent operons. Since operons A and B together represent only a very small part of the overall sporulation sequence, the phenotype resulting from a G_3 mutation might be very difficult to distinguish from that resulting from a mutation in G_1 or G_2. Thus, mutations producing very similar or identical

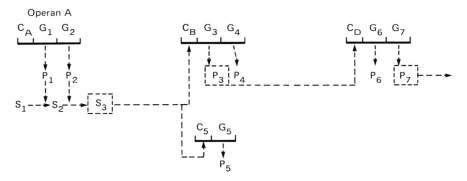

Figure 10-3 Schematic representation of a model describing the regulatory circuit of a sporulation operon in *Bacillus*. There are four operons — A, B, C, and D. Each contains structural genes G_1, G_2, G_3, etc., coding for proteins P_1, P_2, P_3, etc. Where the proteins are enzymes, they act on a substrate S, P_1 on S_1, P_2 on S_2, etc. Each operon also contains a control element (C_A in operon A, C_B in operon B, etc.), controlling its transcription. Compound S_3 acts on C_B and C_C to switch on operons B and C, respectively; protein P_3 acts on C_D to switch on operon D. (Adapted from Piggot et al. 1981)

phenotypes may map quite separately. In contrast to a G_3 mutation, mutation in G_4 of operon B would not prevent the expression of subsequent operons. Mutation in G_3 would be a typical pleiotropic spo mutation, whereas, if P_4 had no regulatory role and functioned solely in the final assembly of a complete spore, then mutation in G_4 would give an incomplete spore—causing, for example, incomplete spore resistance or altered spore germination, or both —and could be represented by a late spo or a ger mutation. Mutation in the controlling element C_b could lead to constitutive functioning of operon B and of other operons whose activity depended solely on operon B; such mutations might be partial supressor mutations or bypass mutations as described earlier. Thus, mutations mapping close together, in C_b, G_3, and G_4, could give rise to very different phenotypes. Operon C is subject to the same regulation as operon B, but maps in a separate location. If its products have no regulatory role in spore formation, then a mutation in G_5 would generate a phenotype similar to one in G_4 and could, for example, give rise to a Ger⁻ phenotype. Thus, genes mapping far apart on the chromosome and represented by mutations generating very different phenotypes (Spo⁻ for G_3 and Ger⁻ for G_5) could still be subject to coordinate regulation.

F. Transposable Elements and Developmental Regulation

Finally, a word must be said about the role in development of gene inversions and transpositions, referred to in general as "recombinational control." The incredibly prescient work of Barbara MacClintock on transposible elements in corn took over 20 years before it found its way into the thinking about variability in bacteria. Then, with explosive rapidity, the notions of insertion sequences, transposons, and invertible sequences found their way into many aspects of bacterial regulation.[19] This kind of control is fundamentally different from the classical models of regulation involving feed-back loops that modify the activity of key proteins or turn their synthesis on or off completely. In recombinational control, a cell is given the option of alternative pathways; it is like a railway switch that allows the train to move from one track to another. It is uniquely useful in that it is an either/or mechanism that does not allow both of two alternative phenotypes to coexist in the cell. I have already discussed the work of Melvin Simon's laboratory on the invertible sequences and phase variation in *Salmonella*. At the end of their paper that first described the work, the researchers state that "evidence in a variety of different systems has accumulated suggesting that a recombinational event, an inversion . . . or a transposition . . . can be involved in regulation of gene expression. These findings taken together with our present view of phase variation suggest that it may be worthwhile to reexamine the notion that directed recombinational events play a role in cell differentiation and development."[20]

G. Epilogue

The primary processes of development—cell division, cell movement, cell adhesion, cell differentiation and cell death—defy analysis in terms of lists of genes and linear interactions of their protein products. Each of these processes is the result of myriads of molecular events which are regulated in parallel not only by interacting genes but also by concomitant epigenetic events: Those not specified directly by particular genes. Given the microscopic nature of all these events and their remarkable complexity, it is perhaps no wonder that there is as yet no adequate theory of development in the same sense that there are adequate theories of genetics and evolution. Constructing such a theory is a particularly compelling challenge because the determination of the nature and regulation of genetic and epigenetic sequences in embryonic development and their relation to evolution is perhaps the grandest outstanding puzzle in biology.[21]

Chapter 11

Exchange of Signals

> "Ye blessed creatures, I have heard
> The call ye to each other make.
> —Wordsworth, "Ode: Intimations of Immortality"

A. Introduction

The exchange of signals between cells necessarily implies some sort of heterogeneity, which may take a number of forms. The population exchanging the signals may comprise different cell types, and that heterogeneity may be a relatively stable, genetic one. One such example is the signaling between opposite mating types found among the aquatic water molds.[1] Alternatively, the population may comprise genetically identical but functionally differentiated cells such as would be found in a complex multicellular organism. The signaling role of plant and animal hormones is an excellent example of such an arrangement. Finally, one could imagine a population of genetically and functionally identical cells within which social or interactive phenomena occur and where signaling is involved. If such a population had to construct and maintain a geographical asymmetry, some sort of cell-to-cell communication would be necessary. This would be the case whether the heterogeneity involved calling the cells in from an otherwise even distribution, so as to aggregate them, or whether it involved the construction of some nonrandom structure.

The bacteria are traditionally thought of as prototypically unicellular creatures. That is to say, the properties of a population of bacteria are no more or no less than the sum of the properties of the individual cells. Yet no one who has watched with amazement as an entire colony of *Bacillus rotans* migrates across the surface of an agar plate or who has examined the behavior and properties of the higher actinomycetes can doubt that these organisms are behaving in some sort of corporate sense. Pattern formation among the heterocyst-forming cyanobacteria almost demands that one consider signal exchange as a coordinating mechanism, and a brief encounter with the myxobacteria is enough to convince the most skeptical that a variety of cell interactions characteristic of a primitive multicellularity exist among the bacteria.

However, there are relatively few instances among the prokaryotes where it is clear that some form of intercellular signaling has actually been demonstrated. These instances include the mating pheromone in *Streptococcus*, the cell density signal for luciferase induction in *Photobacterium*, heterocyst and akinete formation in the cyanobacteria, and various developmental signals in the myxobacteria and *Streptomyces*. Some of these will be discussed in this chapter; however, biologists have been looking at cell-to-cell signaling in eukaryotic microbes for a long time, and we have a great deal to learn from these systems. The following is an attempt to categorize some of the major cell interactions that occur among eukaryotic cells.

I. Signaling that takes place via a diffusible, extracellular molecule.
   A. chemotactic signaling
      1. cAMP as acrasin in Dictyostelium discoideum[2]
      2. leukocyte taxis[3]
   B. neurotransmitters[4]
   C. hormonal effects[5]
   D. mating signals
      1. yeast[6]
      2. *Achlya*[7]

II. Signaling via cell contact interaction
   A. cell recognition, e.g., sponge cells[8]
   B. mating signals, e.g., *Chlamydomonas*[9]
   C. transitions from a unicellular to a multicellular state, e.g., in *D. discoideum*[10]
   D. generation of the immune response[11]

I shall try to set the stage by discussing some of these systems where useful information is available and strategies have emerged. The eukaryotic systems to be discussed with some detail are chemotactic signaling and cell-contact interactions in *D. discoideum*, as well as mating signals in yeast and *Achlya*.

B. Cyclic AMP, Chemotaxis, and Developmental Aggregation in D. discoideum

The aspect of the biology of *D. discoideum* that has made it so interesting to investigators and so pertinent to the general topic of cell interactions is that the organism undergoes a transition from a unicellular to a multicellular state. Thus, one has the opportunity to examine not only the forces that keep cells together in a multicellular entity but also those forces that bring them together after having been widely dispersed. While feeding, the myxamoebae of *D. discoideum* may occupy only about 1% of the total surface area. In order to bring these cells in to focal points over distances as great as 20 mm, it is obvious that there must be some sort of diffusible chemical signal. That signal in *D. discoideum* has been identified as cAMP. As such, it is one member of a general class of attractants referred to as *acrasin* and characteristically functioning as aggregation signals in the Acrasiales or cellular slime molds. The signal was named after the witch Acrasia in Spencer's "Fairie Queene" who has the ability to attract men to her, and like the Greek enchantress Circe, then turns them into animals:

> . . . the faire Witch her selfe now solacing, with a new lover, whom through sorceree and witchcraft, she from farre did thither bring: . . . more subtile web Arachne cannot spin.[12]

In brief outline form, the sequence of events involved in the aggregation process, are first, that the cells are stimulated to begin aggregating when the nutrient level drops below a threshold. Some of the cells respond by excreting cAMP in slow, rhythmic pulses—about 1 pulse per 10 minutes. The rest of the cells produce cAMP receptors (referred to as cAMP sites) on their surfaces. The cells also produce the enzyme cAMP phosphodiesterase, which breaks down the cAMP and prevents it from accumulating. The cells that perceive the pulse of cAMP begin to move in the direction of the source of the signal. (The mechanism whereby they perceive the gradient is not clear, but some evidence suggests that, like bacteria but unlike leukocytes, they use a temporal sensing mechanism.) They move for about 1.5 minutes at a speed of about $12\mu m/minute$. A key feature of the process is that the cells then relay the information; stimulated cells release a puff of cAMP about 10 seconds after they themselves have received the stimulus. They then stop and wait a few minutes until they encounter the next wave of outward-bound attractant. Thus, the signal is effectively amplified and propagated. Part of the aggregation process is pictured in Figure 11-1. The moving and stationary cells have different optical properties; thus, the periodic movement results in alternating dark and light bands of cells. This striking pattern is illustrated in Figure 11-2.

An important part of the transition from unicellular amoebae to a multicellular pseudoplasmodium is that the cells become adhesive. This aspect is

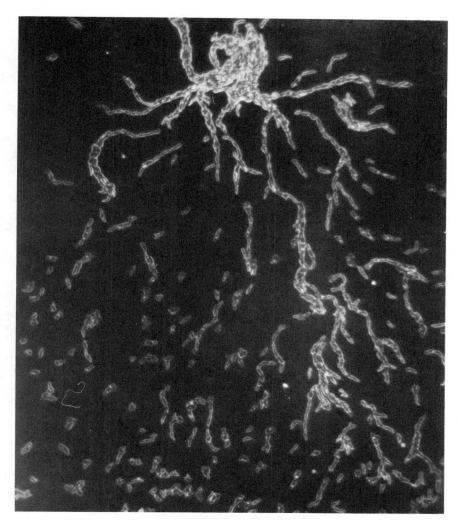

Figure 11-1 Amoebae of *Dictyostelium discoideum* aggregating on a thin layer of agar. The individual amoebae are about 10 μm in diameter. Picture taken using dark-field microscopy. (Courtesy of Dr. P. Newell).

not merely a reflection of some sort of nonspecific stickiness but rather of the fact that specific macromolecules on the cell surface enter into intercellular interactions with each other. Such interactions are the basis of multicellularity and as such are the focus of intense interest. One important aspect of this work was initiated and largely developed in the laboratory of Gunther Gerisch. In *D. discoideum*, two classes of cell-surface macromolecules have been shown to play a role in direct cell-to-cell interactions. During vegetative growth, cells are capable of a side-by-side adhesion. This

Figure 11-2 Amoebae of *Dictyostelium discoideum* showing spiral and concentric waves of cells that are the precursors to the streams of cells in Figure 11-1. The alternating waves represent alternating periods of movement and rest. Picture taken using dark-field microscopy. (Courtesy of Dr. P. Newell)

adhesion is disrupted by EDTA. During aggregation, the cells enter into an end-to-end adhesion that is not disrupted by EDTA. It is possible to raise antibodies against membranes of cells from both these stages. If antiserum against aggregating cells is adsorbed with antiserum from growing cells, the antiserum is thus made essentially specific for aggregating cells. Conversely, the antiserum raised against growing cells contains no antibodies specifically against developing cells, as the corresponding antigens have not yet been synthesized. Thus, one has antisera directed against the cell surfaces of growing and developing cells. These antisera were treated proteolytically to generate univalent antibodies referred to as Fab. These antibodies have only a single reactive site; thus, they can react with a corresponding antigen on a cell surface but are unable to form intercellular lattice connections. As a result, cells thus treated remain in suspension, and any interference with their developmental behavior can be attributed to blockage of cell surface sites rather than nonspecific agglutination.

Using this approach, it has been possible to demonstrate that 2% of the surface of the developing cell is occupied by what have been called "contact sites A." These are sites that, when blocked by Fab specifically directed against them, block the ability of the cells to aggregate. It could be shown that there are about 300,000 such sites evenly distributed along the cell surface. Recently, these sites have been isolated and have been shown to be a single glycoprotein with a molecular weight of 80,000.

More recently, two additional such molecules have been described on the surface of developing cells of *D. discoideum*. One of them is also a glycoprotein with a molecular weight of 95,000 that also participates in cell-to-cell adhesion but at a later stage in the developmental process than does contact site A. The third such molecule has been shown to be involved in the cell-adhesion process. It too has been demonstrated by means of the immunological approach. It was not detected as part of the previous such investigations because the molecule is also present on vegetative amoebae; and in these previous investigations the antiserum that was raised against developing cells was adsorbed with vegetative amoebae so as to leave only antibodies specifically directed against the developing cells. (This illustrates a common attitude that, for a developmental biologist, the only molecules worth studying are those specifically and exclusively synthesized during development. This attitude reflects the view that the only important regulated events are those that turn the gene expression on and off during development. This is, of course, an unnecessarily narrow view of developmental regulation.) Returning to cell adhesion, the third adhesion molecule accumulated during development and seems to be a high molecular weight polysaccharide. Its presence on vegetative amoebae may indicate that, during development, it is somehow modified and changes its location or orientation or the ligand with which it interacts changes. In any case, it is clear that there is a class of macromolecules on the surface of *D. discoideum* that are involved in cell adhesion, that play a role in development, and that characteristically contain sugar residues. It has been suggested that these molecules do actually function to bind the cells together during the conversion from the unicellular to the multicellular state. If so, it is likely that the carbohydrate portions of the molecules play a central role; it is not clear, however, whether similar molecules interact intercellularly with each other or whether there are as yet undefined contact site receptor molecules.

Another class of macromolecules originally thought to have been involved in cell adhesion are the slime mold lectins. These were first discovered in 1973[13] and have been referred to as *discoidins*. They were found by virtue of their ability to cause the agglutination of formalin-treated sheep erythrocytes and were isolated on the basis of their ability to bind specifically to galactose residues. Discoidin has been resolved into two distinct molecules: discoidin I is a tetramer with subunits of molecular weight of 76,000; discoidin II is also multimeric with subunits of 25,000 molecular weight. The two proteins differ substantially; they seem to be different gene products. Even though they are lectins, neither of them has any sugar moiety in the molecule. It has been suggested that they have one or both of two functions. First, that they play a role (as is the case for contact sites A) in the adhesion portion of the unicellular-multicellular transition. In support of this position, a mutant with an altered carbohydrate-binding site on its discoidin I was also blocked in its ability to form aggregates. Here it is extremely important to emphasize that most mutants of *D. discoideum* have been heavily mutagenized. Furthermore, they have usually not been genetically purified so as to make them otherwise isogenic with the parent strain. Thus, unless it is otherwise clear, any such mutant must be viewed as being genetically riddled with multiple mutations. Of course, this makes it virtually impossible causally to associate two properties with each other by determining whether or not both of these properties are altered in the mutant strain. On the other hand, if a mutant has clearly lost a property and is still able to carry out a particular process, one may tentatively conclude, with some confidence in the logic, that those two properties are not causally related. In this sense, a discoidin-minus mutant has, in fact, been isolated and has been shown to be capable of apparently normal development. Finally, a careful attempt to demonstrate a cell-surface location for discoidin has been unsuccessful. Thus, at this point, one must conclude that it has not been demonstrated that discoidin plays any role in the developmental interactions involved in *D. discoideum* development.[13]

The two classes of adhesion molecules, contact sites A and slime mold lectins, do not interact with each other. In other words, the contact sites A are not the receptor sites for the lectins. What roles they play and how these roles are orchestrated with the fascinating transition from the single celled to the multicelled state remains a mystery.

C. Mating Signals in Yeast

It was recognized quite some time ago that the process of sexual mating in the yeast *Saccharomyces* involved the exchange of signals. The suggestion was made that these signals be referred to as *erogens*.[14] The term, however, despite its combination of poetry and anthropomorphism, never caught on. Instead, the signals are commonly referred to as "pheromones" in obvious analogy with the sex attractants of insects. It was believed that the signals are a necessary prelude to mating—coyly referred to as *foreplay* by one author. The term *foreplay* turned out to be more apt than anticipated, for the signaling appears to be commonly associated with the early stages of mating and increases the likelihood that mating will take place; but it is not required. Mutants deficient in various aspects of the signaling process can, under the right conditions, complete the mating process anyway.

Yeast mating is part of one aspect of the yeast life cycle. It results in the formation of a diploid zygote, which can then enter the process of sporulation. The spores are the result of meiotic divisions, show a classical Mendelian segregation of genetic properties, and give rise to the haploid cells. The process is diagramatically illustrated in Figure 11-3. Generally in *Saccharomyces cerevisiae* there are two mating types, referred to as "a" and α. In the most simple of the possible interactions, if "a" and α cell types are mixed with each other, each of the haploid mating types excretes an oligopeptide pheromone that induces the other type to undergo three changes: (1) growth is arrested at the G-1 phase—that is, shortly after bud formation; (2) the cell surface changes in such a way that cell-to-cell adhesion is strengthened; and (3) cell wall and membrane alteration occur that are more appropriate for cell fusion than for growth and budding.

The α factor is produced constitutively and has been shown to be a linear tridecapeptide with the following structure:[15]

$_2$HN-Tryp-his-tryp-leu-gln-leu-lys-pro-gly-gln-pro-met-tyrCOOH

The amino acids are unmodified, the ends are unblocked, and the molecule is extremely hydrophobic. Synthetic molecules will elicit exactly the same response from target cells providing unambiguous confirmation of the nature of the natural α factor. "A" factor has also been isolated and has been shown to be a highly hydrophobic undecapeptide. Studies of the kinetics of binding have suggested that there are specific receptor sites on the cell surface for "a" and α factors. However, these sites have not been isolated or characterized. Some changes in the nature of the surface polysaccharides of the cell have been described; however, there is no clear-cut evidence that any of these changes are associated with the mating-specific agglutinability that develops in the cells. Nor has it been possible to isolate surface molecules, macro or otherwise, that specifically block or interfere with mating adhesion. In fact, there is little if anything to add to this fascinating story regarding how the signal is received or is actually transduced into a cellular response.

Nevertheless, one of the most exciting stories that has unfolded in microbial development has concerned the regulation of the switch that determines the a/α mating type[16] It turns out that once again the switch is a genetic

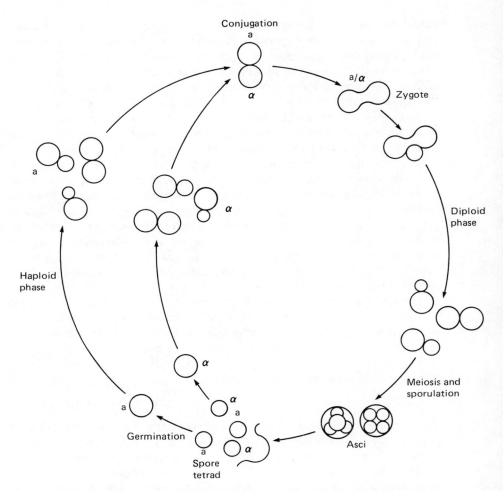

Figure 11-3 Diagram of the life cycle of the yeast *Saccharomyces cerevisiae* depicting the developmental alternatives open to a heterothallic strain (*a* and *α* are the two mating types). (Adapted from J. Thorner 1980)

rearrangement as in *Salmonella* phase variation and G-loop inversion in phage mu (see Chapter 2). The nature of the rearrangement, however, turns out to be quite different. The proposed regulatory mechanism has come to be referred to as the "cassette model" and has emerged largely as the result of the work of Ira Herskowitz's group. The model, briefly, proposes that the genetic information for the "a" and *α* mating types is contained in genes that are transportable from a storage site to one adjacent to the appropriate promoter. When the "a" gene is in place, it is transcribed. When it is removed, its place is taken by an *α* gene, which is then expressed. There, the cell has the option of being either "a" or α phenotype, with some regulated frequency. The result of this is that the population of cells will be assured of containing both mating types at any time. How is this all done? First, let us consider the relevant observations:

1. In order for mating to occur, there must be two mating types present, namely, "a" and α. These are referred to as MATa or MATα.

2. Among some strains, referred to as *heterothallic*, a population of "a" or α cells will spontaneously generate a high proportion of the opposite mating type α (or "a") cells during vegetative growth of the haploid cells.

3. Whether or not the culture is capable of this behavior or, in contrast, is a stable mating type (homothallic) depends on the presence of a single genetic determinant HO, present in heterothallic strains.

4. Cells with a mutation in the mating type locus have the surprising ability to phenotypically heal MATa or MATα mutations. In other words, a MATα cell was still able to mate with a homothallic MATa cell. The progeny are MATα.

5. It could be shown that the repair depended on the presence of the HO locus.

One possible explanation of mating type interconversion is the same "flip-flop" model that applies to phase variation in *Salmonella*. You may recall that this is based on the ability of a piece of the genome to reverse its orientation so that transcription is either permitted or prevented. While this model would explain the interconversion between MATa and MATα, it could not readily explain the healing of the MAT mutations. Instead, it has been proposed that the cells contain at least one extra MATa and MATα locus at sites other than the allele present in the actual mating type locus. These extra MATa and MATα genes are not expressed but when transported to the actual site of the mating type locus are then transcribed. A considerable amount of genetic evidence for this cassette model has been obtained, and many of the interesting details are now available. For example, the transport of a MAT gene from its storage site to the site of expression does not involve its excision from the storage site but rather its duplication. Of course, in contrast to the flip-flop model the MATa and MATα loci are nonidentical sequences of DNA, and the process of cassette replacement is, in some way, dependent on some product of the HO gene. Furthermore, genetic analysis of MAT mutants has shown that the MATα locus is a complex one, containing at least two separate functions that have been designated MATα1 and MATα2. It has been suggested that MATα1 turns on α-specific functions and that MATα2 turns off "a"-specific functions; "a" is otherwise constitutive.

The model has been likened to a computer where stored information is transferred to a display register, and the previously displayed information is erased upon removal. It has been proposed that the model is applicable to the developmental generation of multiple cell types in higher eukaryotes. Thus, for example, if instead of MATa and MATα there were cassette units for generating cell types, A, B, C, . . ., and these could be placed in a playback locus when needed, the necessary multiple cell types could be generated.

D. Steroid Hormones and the Water Molds

The last area of eukaryotic signal exchange in microbes that I want to discuss is the exchange of mating signals in the water mold *Achlya*. The great modern mycologist, John Raper, originally showed that mating in this genus was initiated in response to the production of diffusible substances reciprocally excreted by the two mating partners.

Achlya is a mycelial fungus that lives in aquatic environments. It belongs to the class of fungi called Oomycetes, which, based on metabolic pathways, molecular weight of ribosomal RNA, cell-wall chemistry, and composition of chromosomal proteins, represent an evolutionarily unique group of fungi. They are further distinguished by two features that make then particularly useful and interesting for developmental biologists: (1) they have the fastest known growth rate for a eukaryote, and (2) they are the most primitive eukaryotic organism to produce and respond developmentally to steroid hormones.[17] The situation is as follows. Normally, *Achlya* is heterothallic, that is, male and female strains are separated. The female strain continuously produces a steroid hormone called *antheridiol*. Its structure is:

The antheridiol induces the male strain to produce the male sexual organs referred to as antheridiol initials.

The male thallus, thus stimulated, produces and excretes its own hormone, called *oogoniol*. This induces the female to form the oogonial initials, the female sexual organs. Oogoniol is also a steroid, and its structure is:

[Chemical structure of oogoniol]

The male initials apparently perceive the antheridiol and respond chemotropically, eventually fusing with the female thallus. The nuclei within the male antheridia have, by now, undergone meiotic division to form the gametes. These fuse with the oospheres formed by meiotic division of the female nuclei and the zygotic oospores are formed. These will germinate, under the proper conditions, to form the recombinant diploid mycelium. The life cycle is illustrated in Figure 11-4.

A fair amount of information, much of which has emerged from Paul Horgen's laboratory, has begun to reveal the biochemical details of the mechanism of the antheridiol effect. The first effect is an increase in the rate of net synthesis of ribosomal RNA and ribosomes. This increase takes place within 30 minutes after the cells are exposed to the hormone. At about 3 hours, there is an increase in the rate of acetylation of the histonelike basic nuclear proteins followed shortly by an increase in the rate of synthesis of polyadenylated mRNA.

It must be obvious that this is a rich system for examining the details of the biochemical basis of steroid hormone functions, the nature of the receptors for the hormones, and in general the molecular details of the whole panoply of the complex eukaryotic response to steroid hormones.

E. Cell Signaling among the Prokaryotes

As indicated at the beginning of this chapter, the systems among the prokaryotes where some sort of signaling has actually been demonstrated and which seem to be most promising as experimental systems, are the mating

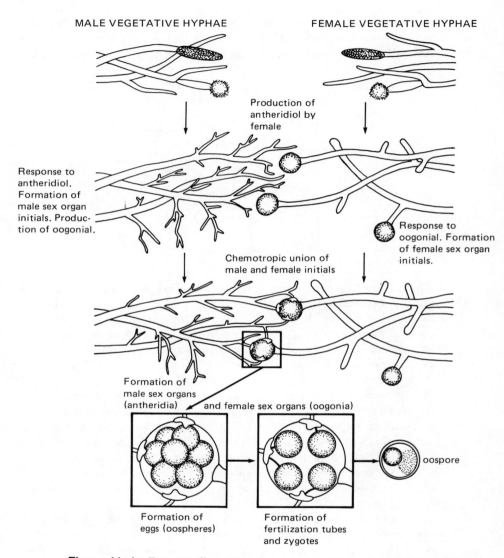

Figure 11-4 Diagram illustrating the exchange of sexual mating signals during the life cycle of the water mold *Achyla*. (Adapted from Alexopoulos, 1962)

signals in *Streptococcus faecalis*, the cell density signals for luminescence in *Photobacterium*, and the developmental signals in myxobacteria and *Streptomyces* (the *Streptomyces* signals have already been discussed in Chapter 6).

In order to be able to conclude with complete confidence that cell-to-cell signaling is involved in a process, one must demonstrate that:

1. The signal is external to the cell as a normal part of the process.
2. There is a receptor for the signal.
3. When the receptor is occupied by the signal, the response in question occurs.
4. A mutant deficient in some aspect of the signaling is unable to complete the response in question.

These criteria have not been fulfilled for any of the prokaryotic systems mentioned above (nor for most of the eukaryotic systems). However, the evidence is sufficiently convincing to justify the hypothesis that bona fide signaling is indeed taking place.

1. Streptococcal Pheromones

S. faecalis appears to indulge in a similar kind of "foreplay" described earlier as a prelude to sexual mating in yeast. Like yeast, which are also nonmotile, the likelihood of potential partners being thrown together is considerably increased by the induction of clumping. Thus, strains of *S. faecalis* that are to be the recipients of a plasmid from a donor strain produce and excrete a low molecular weight peptide (about 700 to 800 daltons). This factor (referred to as CIA for "clumping inducing agent") has been designated by the author as a pheromone, analogous to the sex pheromone involved in the mating of insects. This peptide causes the mixture of donor and recipient to form clumps and stimulates the transfer of plasmid from donor to recipient. This stimulation results not only from the clumping but the transfer process itself seems to be enhanced by exposure to the factor. Upon receiving the plasmid, the recipient cell proceeds to shut down export of the CIA or to export modified and inactive CIA. It has also been shown that cells of *S. faecalis* produce a variety of CIAs, each specific for inducing the transfer of a particular plasmid; when that plasmid has been transferred, the corresponding CIA is no longer excreted. There is no information as to the specific physiological changes induced by the pheromone, nor as to the pheromone receptor sites.[18]

2. Autoinducer of Bacterial Bioluminescence

Surprisingly, the most thoroughly described cell-to-cell signaling system in bacteria is in a nondevelopmental system, bioluminescence.[19] The function of this excreted autoinducer is to serve as a parameter of cell density. It is analogous, in that sense, to adenosine and orthophosphate, excreted as cell density parameters by *Myxococcus xanthus* (see Chapter 7). The biological significance of such a process is presumably to withhold luminescence until there is a sufficiently high cell density for the light to be perceived by another organism. The cell density effect has been clearly demonstrated with chemostat cultures of a number of different species of luminescent bacteria (for example, *Photobacterium fischeri* and *Vibrio harveyi*). The autoinducer for *P. fischeri* has been isolated and identified as (β-ketocaproyl) homoserine lactone:

$$CH_3-CH_2-CH_2-\underset{\underset{O}{\|}}{C}-CH_2-\underset{\underset{O}{\|}}{C}-NH-CH\underset{\underset{\underset{O}{\|}}{C}}{\overset{CH_2-CH_2}{\diagup}}O$$

It is interesting that, while the autoinducer of one species of luminescent bacteria will not turn on luminescence in another species, nonluminescent species will produce autoinducers for luminescent species in the same genus. (Production of a luminescence autoinducer is not an invariable feature. There are some luminescent bacteria that glow at low cell densities. It would be interesting if this trait was a reflection of a different ecological niche or a different function for such luminescence.)

3. Signals in Myxobacteria

The myxobacteria manifest the most complex and varied signaling among the bacteria. As pointed out earlier, their need to optimize feeding by maintaining a high population density results in a set of social interactions that permeate growth, motility, and development. Their communication is accomplished both by diffusible extracellular signals as well as by contact interactions between cell surfaces. None of these interactions has been thoroughly characterized; however, they offer the opportunity to use a variety of experimental approaches to examine a number of different kinds of interactions. At this moment, physiology, biochemistry, and genetic and molecular biology are coming together to focus on these phenomena.[20] It is an exciting time.

The interactions that seem likely systems for examining cell-to-cell signaling are:

    a. social ("S") motility
    b. extracellular complementation
    c. cell density effects
        1. adenosine and fruiting
        2. orthophosphate and myxospore germination
    d. myxobacterial hemogglutinin
    e. *Stigmatella* pheromone

a. *Social ("S") motility.* The interactive nature of S motility in *M. xanthus* has been referred to in Chapter 7. It has recently been shown that S motility is highly correlated with the presence of polar pili on the cells.[21] Furthermore, when conditional S-mutants are stimulated to move by physical contact with a complementing S-mutant, the transiently motile cells synthesize pili for a brief period of time. It is not unlikely that there is a pilus-pilus or a pilus-receptor association that coordinates the cooperative movement of the cells. Such pilus-receptor associations are not unique; they have been shown to play a role (albeit undefined) in the formation of mating aggregates in *E. coli*.[22] However, neither in *E. coli* mating nor in *M. xanthus* S motility is the role of the pilus, the nature of the pilus receptor, or the precise function of the pilus understood.

b. *Extracellular complementation.* In Chapter 7, I described a series of conditional mutants of *M. xanthus* that were blocked in their ability to form fruiting bodies and myxospores. If a mutant was mixed either with wild-type cells or with mutants from another of the four complementation groups, it was able to complete its development. It was concluded from these experiments that there were at least four signals necessary for and exchanged during normal development. It has been shown that one class of these mutations can be rescued by a mixture of components of peptidoglycan. Another mutant can be rescued by a cell-bound molecule that appears to be a small polysaccharide or glycopeptide. In the latter case complementation requires cell-to-cell contact and does not involve a diffusible signal. The questions one would like to answer are:

- What is the chemical nature of the signal?
- What is its structural/functional relationship with the cell surface?
- What is the nature of the signal receptor?
- How are the activities of the signal and receptor regulated?
- Is it their biosynthesis that is controlled or their accessibility?
- What is the nature of the signal-receptor interaction?
- How is this interaction transduced into a developmental response?

c. *Cell density effects.* There are three aspects of the biology of *M. xanthus* that involve cell density effects. The first is the dependence of the growth rate on cell density when that growth is based on the extracellular hydrolysis of macromolecules.[23] The second is the germination of myxospores, and the third is fruiting-body formation. The first case does not involve cell-to-cell signaling; however, it does in a sense involve cell interactions. And the requirement for a high cell density in order to optimize feeding provides, in my opinion, is the raison d'être for the entire myxobacterial life cycle. The requirement for a high cell density in order for myxospores to germinate has only been demonstrated for glycerol induced myxospores. The spores excrete orthophosphate, and when the external concentration reaches a threshold level, the cells will germinate in distilled water. Spores at a low cell density will germinate if they are in nutrient medium or, if in distilled water, when orthophosphate or a concentrate of conditioned medium is added.[24] Using the external concentration of an excreted molecule is a simple yet cunning device whereby a cell can determine its own cell density. Nothing is known in detail about the process—how the orthophosphate is secreted, what the cellular reservoir of orthophosphate is, how the phosphate is perceived, how its concentration is determined, and how that becomes the signal for germination.

In a similar fashion, cells will not aggregate and form fruiting bodies when placed in an appropriate agar surface at less than a critical cell density. It has been shown that the cells use the same strategy for measuring cell density, except in this case the signaling molecule is adenosine.[25] In what is getting to sound like a litany, nothing is known about the source of the adenosine, the adenosine receptor, and so on.

d. *Myxobacterial hemagglutinin (MBHA).* It may be overly optimistic to include this in the category of cell signaling for there is even less known of MBHA as a signal than any of the other signals discussed. While the MBHA itself has been carefully characterized (see Chapter 7), its role is only surmised. It will bind to the oligosaccharide fetuin, but no myxobacterial ligand has been shown to react with MBHA. It appears during development and then disappears and is localized at the poles of the cell.[26] Thus, it is easy to suggest that it plays a role of some kind in the process of aggregation. This has not, however, been determined.

e. *Stigmatella pheromone.* The observation that provided the clue that a diffusible material was playing a role in the development of *Stigmatella aurantiaca* was that populations undergoing fruiting-body formation induced neighboring low-density populations, which would ordinarily not have fruited, to form normal fruiting bodies. Since then, it has been possible to isolate the inducing material and to obtain some information as to its chemical composition. It is a nonvolatile, low molecular weight material (1800 daltons), soluble in chloroform and resistant to autoclaving and proteolytic digestion. The material has been designated as a pheromone, using the broadest definition of the term. Its effect on the cells is not only to bypass the need for a high cell density but also to spare the light effect. Normally, *S. aurantiaca* requires visible light for fruiting-body formation; in the presence of the pheromone, cells will fruit in the dark.[27] The *Stigmatella* pheromone is the only diffusible, intercellular signal that has thus far been isolated in the myxobacteria.

F. Conclusions

There is a vast literature concerned with the exchange of signals between cells. Almost all of it is with eukaryotic cells and most of it involves the exchange of diffusible, extracellular signals. Two areas of cell signaling have thus far resisted analysis. One of these is the mechanism whereby signal reception is transduced to a response—either biochemical or behavioral. The other is the nature and precise mechanism of cell surface interactions. In most cases, attempts to penetrate into these problems are hindered by the complexity of the systems and the intrinsic difficulty in doing genetics and molecular biology with the system that shows cell interactions. It is now clear that there are well-defined signaling systems in the prokaryotes— autoinducer in luminescent bacteria, mating pheromene in *S. faecalis*, and developmental signals in the myxobacteria. The latter include both diffusible and cell surface signals. These systems offer the realistic possibilities of arriving at a mechanistic description of all aspects of the process—that is, nature of the signal, regulation of signal biosynthesis and export, signal perception, and transduction of that perception to a behavioral or biochemical response. In addition, the existence of a developmental mutant of *M. xanthus* that can produce the appropriate cell surface signal, but is deficient, either in its ability to export the signal or to orient it properly on the cell surface, will be a very nice model system for studying cell contact-mediated interactions.

Chapter 12

Genetic Approaches to Studying Development

A. Introduction

In a very thoughtful and useful review on the subject of the genetic analysis of microbial development,[1] Botstein and Maurer have pointed out that development consists of a series of ordered processes—morphological change, defined sequences of events, formation and dissolution of subcellular assemblies, and directed changes in cell function.

There are a variety of genetic techniques that have been designed or have evolved to ask particular kinds of questions about these processes. And it is a feature of the peculiar genius of genetics that it can produce clues that lead to otherwise obscure or invisible biochemical processes. One need only to recall the genetic experiments on the lac operon that led inexorably to the idea of, and eventually to the isolation of, the lac repressor.

The format I shall use for this chapter is to ask, what sorts of biological/developmental questions can be answered by genetic analysis? And with this as a framework, describe the techniques with which the question can be approached. Where possible, examples in prokaryotic development will be referred to.

B. The Various Approaches

1. Is There a Particular Causal Relationship between Two Properties of a Cell?

The use of mutants is a powerful way to reveal relationships, and the logic is simple. If A and B are causally related, and either is mutated, some response relating to the other should be noted. Many examples leap to mind, but one particularly dramatic example with which I am personally familiar is Roger Stanier's discovery of the photoprotective role of carotenoid pigments.[2] He had obtained a series of mutants of the photosynthetic bacterium, *Rhodopseudomonas spheroides*, that were blocked at various stages in the synthesis of colored carotenoids. Whereas these mutants were obtained for the purpose of studying the biosynthetic pathway of carotenoids, Stanier noticed that one mutant, completely unable to synthesize colored carotenoids, was killed by exposure to visible light and oxygen. On the basis of this unexpected correlation and subsequent experiments, Stanier formulated the theory of the photoprotective function of carotenoids which proposed that the essential biological function of carotenoid pigments is to serve as a protective photochemical buffer against lethal protosensitization by such pigments as cholophyll. Thus, in its simplest form, this approach can either confirm suspected causal relations or reveal relations that are otherwise invisible. There are, however, alternative explanations that must be disposed of. For example, one must be sure that there are not separate mutations in A and B—that is, that the A mutation is indeed a single-site mutation. This can be accomplished either by backcrossing the mutant (e.g., via transducing phage) into the wild type and selecting for the mutant phenotype or by obtaining back mutants to property A and determining whether property B is simultaneously lost.

From a developmental point of view, an excellent example of the value of this sort of approach has been the isolation of antibiotic-resistant mutants that are also developmentally defective. (This strategy has already been described in Chapter 10 for *Bacillus subtilis* and *Myxococcus xanthus*.) For example, in order to determine if transcriptional regulation plays a role in the development process, mutants with an altered RNA polymerase were isolated based on their resistance to the antibiotic rifampin. Many of these were then also shown to be developmentally aberrant. The same approach has been used to demonstrate the involvement of translational control in sporulation by isolating erythromycin-resistant mutants (see Chapter 3).

It would be superfluous to write at length about the techniques of mutagenesis; there are numerous reviews on the subject[3] and the reader is directed to those. It is useful, however, to point out the use of transposons as mutagens.[4] These are relatively small pieces of DNA, flanked by insertion sequences and carrying a drug-resistance gene; they are portable genetic markers that can be inserted into cells, usually via phage infection, and will integrate randomly into the host chromosome. At the point of integration, the interrupted gene is, of course, mutated. In addition, the mutation is selectable owing to the drug-resistance gene. Thus, otherwise nonselectable mutations (e.g., many developmental genes) become in effect, selectable. As indicated earlier, it is now possible to put transposons into *Bacillus, M. xanthus* and *Caulobacter,* thus, in one step, immensely increasing the feasibility of mapping, three-factor crosses, strain construction, complementation, and a variety of other genetic techniques. One cautionary note about the use of transposons is that their effect is polar; if they are inserted in an operon, transcription downstream from the transposon is also interrupted.

2. The Use of Mutants to Reveal Processes That Are Otherwise Invisible or Obscure

I can think of at least three examples of how a mutant can be used to reveal an aspect of a process that would be otherwise difficult or impossible to detect:

a. A behavior such as motility or chemotaxis (or development) is obviously the net result of a number of individual pathways that blend together. Mutants can help to separate these from each other and to reveal the individual threads in the fabric. An excellent example is the work on the control of gliding motility in *M. xanthus.* Simply looking at the behavior of cells during gliding, one might only conclude that the cells sometimes moved as single individuals and sometimes as groups. However, after a number of laboratories had had difficulty in isolating motility mutants, it became clear that there were two separate regulatory systems, the "A" system (for *adventurous*) and the "S" system (for *social*) (see Chapter 7). Mutants in the A system could only move in groups of cells and mutants in the S system preferred to move as single cells (Figure 12-1). Even more interesting, some of the mutants were conditional in that they could temporarily regain motility if placed in physical contact with a mutant of another complementation group. The S⁺ cells were shown to be piliated, the S⁻ cells were not. And when an S⁻ cell was phenotypically complemented by another cell to regain motility temporarily, it also temporarily regained the ability to synthesize pili. Think about the variety of problems and phenomena revealed by this simple set of mutants—enough for a dozen Ph.D. dissertations.

Chapter 12 Genetic Approaches to Studying Development   207

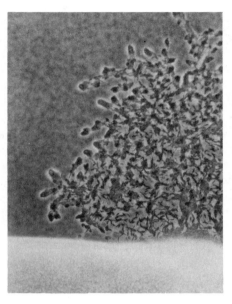
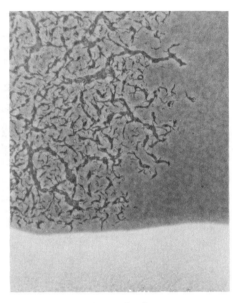

S-motile flare   A-motile flare

Figure 12-1 S motility and A motility of vegetative cells of *Myxococcus xanthus*. The S motile flare illustrates the motility characteristic of an A⁻ mutant; the A⁻ motile flare illustrates the motility of an S⁻ mutant. (Hodgkin and Kaiser 1979)

b. The ability to "rescue" a mutant (which thereupon becomes a conditional mutant) may reveal something about the process that is ordinarily invisible. Obviously, a mutant may be blocked in its ability to make something. If one can then add that "something" back to the cell and thereby correct the mutation, two things have been accomplished. First, one has shown that a particular biochemical intermediate is involved in the process being studied. This can lead to insights into the mechanism of the process itself. Second, by identifying a biochemical intermediate that is involved in the developmental process, one now can ask questions about the regulation of synthesis of a particular gene product rather than attempting to study the regulation of some vaguely defined developmental process.

An excellent example of this approach has come out of Dale Kaiser's laboratory. A series of developmental mutants were isolated that had the interesting property of being able phenotypically to complement each other's development. As described earlier (see Chapter 7), these mutants led to the conclusion that a series of signals were being exchanged between cells as part of the coordination of normal multicellular development. It is hard to imagine how else one could have demonstrated this sort of signal exchange without the use of these mutants.

c. The use of mutants allows the accumulation of biosynthetic intermediates. If a cell is blocked in its ability to complete a particular biochemical pathway, one of two things will happen. Either the accumulated intermediate will feed back and turn off its own synthesis or it will accumulate to the extent that it is identifiable as an intermediate. This approach has been used with immense success in the studies of phage morphogenesis where a large number of phage mutants blocked at various stages of phage assembly have allowed the identification of essentially all of the proteins synthesized during the process of phage head, tail, and baseplate assembly (see Chapter 2).

3. How Many Genes Are Involved in the Process?

Having this sort of information allows one to make some sort of estimate of the complexity of the process. It is obviously useful to know whether the process one is studying involves two genes or twenty genes. This is generally accomplished by determining the number of mutations required to saturate a process. Specifically, one would obtain a large number of independent mutants of the process and then divide them into two categories—those mutations that fall within the same gene and those that are in different genes. This distinction would be made by means of complementation analysis—a method whereby the mutant genes from two different mutants are placed in a single cell. If the mutations are in different genes the combined transcription will yield one complete set of mRNAs and the process will take place in a normal fashion. If the mutations are in the same gene this will not occur—that is, complementation will not take place. Complementation can be accomplished either by using a lysogenic phage to coinsert the mutant gene or by using stable plasmids to generate the partial diploid. The results of the complementation analysis can be corroborated by determining the actual map location of the mutant genes. This would tend to eliminate the possibility of two different mutations in the same gene that resulted in intragenic complementation.

4. Which Genetic Loci Are Regulatory and Which Are Structural?

The amount of DNA present in bacteria (and certainly in eukaryotes) seems to exceed the amount necessary to code for the number of cellular proteins detectable by two-dimensional gel electrophoresis. This is especially the case for *M. xanthus*, whose genome is about 25% larger than that of *B. subtilis*. In general, this is not at all surprising in view of the tremendous amount of regulation that the cell must carry out. Thus, a variety of regulatory elements have been described for bacteria—repressors, corepressors, operators, attenuators, promoters, and activators. It is of obvious interest when dealing with a particular mutant to know if that mutation is of a structural or a regulatory gene. Unfortunately, there is no quick and easy test or set of tests with which to make this distinction; however, there are a few generic distinctions that can be made.

Some regulatory mutations are notoriously pleiotropic; for example, the diffusible protein product (repressor) of a repressor gene may control an entire operon or even many operons. Furthermore, some regulatory elements (such as promoters) exert their effect by virtue of their position rather than as a result of a gene product. This is obviously not the case for any structural gene. Thus, if one can demonstrate that a gene's effect is positional, via *cis-trans* dominance tests, then that is evidence that the locus is acting via a diffusible factor, then there must be a target site for that factor, and it is likely that that target site is encoded by another genetic locus. Thus, a different mutation could mimic all or part of the effects of the first mutation or could suppress the effect of the first mutation by allowing the mutated regulatory factor to bind to a modified target site.

5. What Is the Sequence of Events in, and the Structure of, a Pathway?

It is often possible by careful examination of the morphological events during a developmental process to order the sequence of those events. However, that is more difficult when dealing with biochemical events (e.g., DNA synthesis during development) and in some cases may be impossible. Thus, the development of a genetic method by Botstein's group to determine the order of gene function was most welcome.[5] The technique requires being able to use two different methods to block the expression of each of the two genes whose order one is trying to determine. Using conditional mutants that are either heat sensitive or cold sensitive has thus far been the method of choice. The logic of the experiment is that if genes A and B are expressed in that order, in a linear dependent pathway, and if gene A is cold sensitive while gene B is heat sensitive, then applying a restrictive high temperature followed by a shift to a restrictive low temperature should allow the sequential expressions of both A and B. Reversing the sequence of temperature shifts, or if the gene sequence is B, A, should result in no expression. Furthermore, if both sequences of temperature shifts (hot to cold and cold to hot) allow expression of the pathway, then the A and B mutations are on independent pathways. Finally, if A and B are on separate but interdependent branches of a pathway, then either shift should prevent expression of the pathway. These results are illustrated in Table 12-1 as part of an analysis of the yeast cell cycle.

This ingenious method is useful for determining not only the sequence of genes in a pathway but also for providing information as to the structure of the pathway. This method has been used both in determining the sequence of events during phage (P22) morphogenesis and in the yeast cell cycle.

Another technique for determining whether or not a pathway is a linear dependent one or is branched is called *epistasis*. This is based on the notion that if two genes are part of a linear, dependent sequence of events, a double mutant should have the phenotype of the mutant occurring earlier in the pathway. This technique may be complicated by a number of variables, such as whether one or both of the genes is a regulatory element. However, it has been used successfully to determine the nature of some parts of the pathway of *Bacillus* sporulation.

An excellent example of the sort of process that would be amenable to studies of epistasis and gene sequence is the process of fruiting-body formation in *M. xanthus*. There, at least four major events are intertwined in a way as yet not understood—fruiting-body formation itself, aggregation, developmental lysis, and sporulation.

Table 12-1 Results expected in reciprocal shift experiments using cs ts double mutants

Dependency relationship	Result of shift	
	$17°-37°$	$37°-17°$
Dependent ts→cs→	+[a]	−[a]
Dependent cs→ts→	−	+
Independent ts→ cs→	+	+
Interdependent (cs, ts)→	−	−

[a] A "+" indicates passage to a second cell cycle (i.e., two arrested cells are found and a "−" indicates arrest in the first cell cycle (i.e., one arrested cell is found).

Source: From Botstein, D., and Maurer, R. 1982. Genetic approaches to the analysis of microbial development. Ann. Rev. Genetics 16:61–83.

6. Are There Assemblies of Interacting Gene Products?

In cases where two molecules or subcellular assemblies interact with each other, certain types of suppressor mutations can be extremely valuable.

Geneticists have long appreciated the power of suppressor mutations, but using them properly entails the ability to carry out some rather sophisticated genetic manipulations; thus, they have not been as widely used as their utility would otherwise dictate. In general, an extragenic suppressor mutation is one that occurs in a gene physically distant from another mutation but which acts to reverse the effect of the first mutation. Thus, it appears as a reversion of the primary mutation, but is more properly referred to as a *pseudo reversion*.

A particularly useful type of suppressor mutation is one referred to as an *interaction suppressor*. If it is the function of a particular gene product to interact with another gene product, one consequence of a mutation will be to alter that gene product so that it can no longer interact properly. An interaction suppressor mutation would then result in a modification of the second reactant so that it can once again interact with the other mutated gene product. An excellent example of this is the isolation of interaction suppressor mutants in the α-tubulin structural gene of *Aspergillus nidulans* that corrects mutations in the β-tubulin structural gene.[6] Another way of stating this is that traditional suppression implies a *functional* interaction between genes, whereas interaction suppression implies a *physical* interaction between gene products. The obvious utility of such mutations is that they allow one to determine that an interaction is indeed taking place and to examine the nature of the interacting component. In order to do this, however, it is necessary to be able to identify the function of the suppressor gene itself. Thus, one must be able to demonstrate that the interaction suppressor, in addition to correcting the primary mutation, confers its own new phenotype on the mutant.

Much of the developmental biology of the myxobacteria, and in the future, of other developing systems, focuses around cell interactions (see Chapter 7). Thus, interaction suppressors will be especially valuable for examining various signal-receptor interactions. In addition, any studies concerned with self-assembly as part of morphogenesis will certainly also benefit from the study of interaction suppressors.

7. Strain Construction

Finally, the ability to move genes from one strain to another allows one to construct strains that are useful either for subsequent genetic analysis or for physiological experiments. For example, as indicated in Section B-1 of this chapter, one may need to clean up a mutated strain so that it contains only the single mutation one is interested in examining. This can be done by back-crossing the desired mutation (e.g., by transduction) into the wild-type parent.

C. Transposons and Recombinant DNA Technology

The brilliant success of modern genetics has been nowhere more evident than in that paragon of genetic versatility, *Escherichia coli*. How many other systems are there that are capable of sexual mating, transformation, transduction, plasmid exchange, possess lysogenic phage—all of which take place in the context of a 15-minute generation time? It is not generally emphasized that the great majority of bacteria are years away from being amenable to these routine bread-and-butter genetic manipulations. However, the *deus ex machina* has dropped on the stage in the shape of recombinant DNA technology and transposon genetics. Now, much of the laborious difficulty of classical genetic manipulation has been reduced by the new genetic technology. It not only circumvents the difficulty in most organisms of moving genes from one cell to another but allows the selection of otherwise nonselectable properties (e.g., developmental processes) as well as in the in vitro modification of isolated genes.

The key elements are relatively few: plasmids, restriction endonucleases and ligases, lysogenic phage, and transposons.

The most useful transposons are pieces of DNA containing one or more drug-resistance genes and flanked by two insertion sequences.[4] These insertion sequences can cause the transposon to be inserted by means of "illegitimate recombination" at almost any site along the genome. This has two main consequences. First, the gene into which the transposon is inserted is interrupted and its function destroyed. Thus, a transposon is, in effect, a mutagen that generates random, polar, null mutants. Second, a selectable marker (drug resistance) is inserted as part of the transposon. Thus, any gene that can be cotransduced with the transposon can be coselected. In other words, if a transposon is inserted next to a developmental gene, infection of that cell with a transducing phage will allow the simultaneous transfer and selection of both the drug-resistance gene and the piggybacked adjacent developmental (or other nonselectable) gene. All that is needed is a way to get the transposon into the cell (usually a phage or a plasmid) and a good transducing phage. These are now available for *M. xanthus*, *Caulobacter*, and *Bacillus*.[8] They have been fairly extensively employed in *M. xanthus*, which can be infected by the coliphage P-1 carrying a Tn5 transposon. A number of laboratories have generated essentially complete libraries of Tn5-bearing strains of *M. xanthus*. In such a library containing a few thousand random, Tn5-bearing strains, it is highly likely that every genetic locus in the *M. xanthus* genome is represented by at least one strain containing a Tn5 transposon in that locus.

Let us say that one wished to have a strain of *M. xanthus* that contained a selectable gene (for example, kanamycin resistance) adjacent to, or at least within cotransduction range of, an otherwise unselectable gene (e.g., one involved in fruiting-body formation). Such a strain would be of obvious and considerable value for any manipulation that required selection of the trait—for example, complementation, mapping, back-crossing. How would one do it? One would first pool all the Tn5-containing strains and infect that pool with a generalized transducing phage. There is a reasonable probability that at least one of the transducing particles would infect a cell in which the Tn5 was adjacent to the desired gene. There is likewise a probability that a subset of that class of phage would have packaged just that piece containing the fragment of the genome with both the Tn5 and the developmental gene. If this mixed phage population is then used to transduce a mutant, deficient in the developmental property desired, one can select for kanamycin-resistant transductants and screen those for the described developmental trait. This will include only those cells that have received, via transducing phage, both the kanamycin-resistance gene and the developmental property. If, from among this group of transductants, those from which the two traits can be cotransduced with high frequency are selected, one then has a strain where the two loci are physically close to each other. This elegant and ingenious approach was used to considerable advantage in Dale Kaiser's laboratory to isolate mutants of *M. xanthus* deficient in their social behavior.[8]

Another aspect of the use of transposons addressed the problem of the expression of developmental genes. Often the expression of such genes is subtle, the gene products are unidentified, or the assay for gene expression is awkward or difficult. By fusing the promoterless structural gene for β-galactosidase *(lacZ)* to a transposon that contains a gene for antibiotic resistance (e.g., kanamycin resistance in Tn5), one can insert the *lacZ* gene into the chromosome or plasmid of a developing cell. The virtue of this approach is that the availability of chromogenic substrates for β-galactosidase allows one to visualize by a simple color change in the colony whether or not the gene is being expressed.[9] This will occur if the *lacZ* transposon happens to be inserted adjacent to a promoter that is being expressed. If the promoter is one that normally regulates a developmental gene, one has a convenient measure of the effect of a variety of conditions on the expression of that gene. This has been accomplished with *M. xanthus* in Kaiser's laboratory (see Chapter 7). In a variation of this approach, it is possible to fuse the *lacZ* gene to a cloned, defined developmental gene and to study its expression in the developmental cell. This approach has been carried out in David Zusman's laboratory with the genes for S protein in *M. xanthus* (see Chapter 7).

There are essentially three aspects to the use of recombinant DNA technology for developmental studies. One refers to the ability to remove a fragment of DNA from the organism being studied, to place it within *E. coli,* and then to use *E. coli,* either to amplify the gene or its gene products, or to study that gene's expression in the context of *E. coli's* regulatory system. Another more interesting aspect is to reinsert a fragment of DNA back into the cell from which it came. Thus, the various developmental consequences of complementation, gene dosage, and localized mutagenesis can be examined. Finally, having reagent amounts of the gene would make it possible to do such things as sequencing and heteroduplex or restriction mapping of the gene.

As examples of the value of cloning the gene in E. coli, I have already described the cloning of various sporulation loci of Bacillus and the use of the cloned fragments to examine the specificity of RNA polymerase (see Chapter 10). In M. xanthus, the gene for the S protein of myxospores has been cloned and has been used as a probe to reveal the unexpected fact that the gene exists as a tandem duplication in the cell.[10]

The utility of being able to reinsert the cloned gene back into its native host is illustrated by recent work that used the coliphage P1 to reinsert a cloned developmental gene back into M. xanthus, whereupon dominance and complementation tests were performed.[11]

The possibilities for using these techniques will be limited only by the imagination and ingenuity of the investigators. The entire spectrum of questions relating to the regulation of specific gene products associated with specific developmental events is now beginning to be available.

D. Conclusions

Among the many kinds of questions that can be addressed using genetic analysis, some of them have developmental relevance, and this chapter has tried to outline and describe them. Among the various techniques that were involved are mutant analysis, complementation, *cis-trans* dominance tests, the use of conditional mutants, epistasis tests, suppressor analysis, gene mapping, and phage transduction. In addition, the use of transposons, lacZ fusions, and recombinant DNA technology has added an entire new dimension to the kinds of questions that can be asked as well as erasing many of the normal genetic barriers between cells.

Four of the developing prokaryotes are amenable to these genetic analyses. These are Bacillus, M. xanthus, Caulobacter, and Streptomyces. The obvious experimental advantages of prokaryotic genetics need not be dwelled upon. The genetics of Bacillus has the longest history and is the most highly advanced having available to it lysogenic phage, generalized and specialized transduction, transformation, plasmids, protoplast fusion, transposon biology, and recombinant DNA technology. It has been limited until recently by the failure to define the developmental processes in terms of specific gene products and, again until recently, by the inability to generate stable partial diploids.

The genetics of *M. xanthus* is still at the stage of taxiing down the runway, but it is picking up speed. There are available lysogenic and generalized transducing phage and transposon biology via infection with coliphage P1, and it is possible to construct partial diploids. There have been no native plasmids demonstrated yet; neither is there a transformation nor mating system available. However, a number of developmental gene products have been described, and, of course, the recombinant DNA and transposon techniques are being vigorously exploited.

Caulobacter genetics is also at a take-off point. Well-defined gene products are available (e.g., flagellin), a transposon (Tn5) can now be inserted into the cell, RP4-mediated conjugation has resulted in a genetic map of *Caulobacter crescentus,* and fine-structure mapping with a generalized transducing phage has been carried out.

Chapter 13

Epilogue

Having reviewed the major experimental systems among developing prokaryotes and discussed the developmental problems surrounding them, it is now appropriate to place them in a larger perspective and to attempt to look ahead into the future of the field.

Developmental biology as a discipline emerged from embryology; and while it has traditionally been focused on the problems characteristic of multicellular organisms—most commonly animals—the past 50 years or so have been the emergence of what might be called developmental microbiology. There were a number of advantages to studying development in microbes. They were experimentally far more convenient than multicellular plants or animals and, in addition, in some of them it was possible to separate growth from development, immensely simplifying the system. Thus, a great deal of work has been done with basidiomycetous fungi such as *Schizophyllum*, the water molds *Achlya* and *Blastocladiella*, green algae such as *Chlamydomonas*, the protozoa *Paramecium*, *Acanthamoeba*, and many others.[1] Among these eukaryotic microbes, however, the most attention has been devoted to the cellular slime mold *Dictyostelium discoideum*. This work originated with Kenneth Raper, was further developed by J. T. Bonner and Maurice Sussman, and now, from a biochemical and molecular point of view, the information pertaining to development in this organism is as sophisticated as in any system available.[2]

Each of the traditional developmental systems—multicellular eukaryotes and eukaryotic microbes—has its own peculiar experimental virtues and shortcomings. For each there is a subset of developmental processes or problems that is either uniquely or most conveniently studied in that system. For example, it is difficult to imagine a better system than the cAMP-mediated developmental aggregation in *D. discoideum* for studying the generation, relaying, perception, and transduction of a signal. Likewise, the transition from the unicellular amoeboid state of *D. discoideum* to the multicellular pseudoplasmodium is admirably suited for examining the nature of multicellular interactions and the transition from the unicellular to the multicellular state.[2] If one is interested in the problems of developmental commitment and multicellular differentiation, then there are not many alternatives better than studying the imaginal discs of *Drosophila*. Nevertheless, I hope there will have emerged from this book the recognition that there is now another phase of developmental biology centered around the developmental behavior of certain bacteria, and it seems clear that, in a similar sense, a subset of developmental problems has emerged that can be effectively, or perhaps even more appropriately, examined in bacteria. Some of these are:

1. *The regulation of expression of developmental genes.* One of the stumbling blocks in the study of eukaryotic development has been the difficulty in obtaining in vitro systems for generating meaningful gene products. Thus, questions of the regulation of gene expression at the molecular levels of transcription and translation have remained unanswered. It is clear that such approaches are now available with all of the prokaryotic systems discussed in this book, and I would expect that at least the sorts of detailed regulatory insights presently available for lambda phage and the lac operon will emerge for developmental genes and operons. Add to this the dimension of movable genetic elements plus some, as yet undiscovered, regulatory strategies, and the prospects are truly exciting.

2. *Nutrient signaling of development.* In the case of the microbe, the relation between an environmental factor and development is usually far more direct than is the case with a multicellular organism. Microbial development is often a response to a nutrient signal; and the detailed examination of the nature and function of second messengers such as cAMP and guanosine tetra- and pentaphosphates has laid a groundwork for approaching the problem of the biochemical transduction of environmental signals to developmental events. Assuming that the model of a second messenger is appropriate, the operative questions then are (a) what is the relation between the nutrient signal and generation of the second messenger, and (b) what role does the second messenger play in regulating differential gene expression? Such developmental processes as endospore and myxospore formation in *Bacillus* and *Myxococcus*, respectively, developmental aggregation in myxobacteria, and heterocyst formation in cyanobacteria are ideal systems for examining these questions:

3. *Cell morphogenesis.* As pointed out in Chapter 9, the question of the chemical determinants of cell form is one that is rarely addressed and for which few insights are available. It has long been my optimistic view that the existence in bacteria of a cell-shaped macromolecule (i.e., peptidoglycan) would provide the basis for a chemical understanding of the determination and maintenance of cell shape. That has not turned out to be the case, perhaps because it is a problem that can only be addressed by adding a three-dimensional view of the process to the more conventional biochemical approaches. Such approaches tend to be taken by physical biochemists who have not yet expressed much interest in the problem. Nevertheless, the process of shape change as a developmental problem remains to be addressed.

4. *Spatial differentiation.* This problem, which concerns the ability to generate an asymmetric cell shape, is prototypically studied in *Caulobacter* and hardly addressed in higher systems. In a sense, it is a subset of the general area of cell morphogenesis. The work of Shockman's laboratory on the regulation of division in the streptococci (see Chapter 9) has laid a conceptual and experimental groundwork for this question in a nondeveloping bacterium. In *Caulobacter,* work on flagellin mRNA segregation, nucleoid-flagellin-membrane interactions, and their relation to the cycles of growth and DNA replication (see Chapter 4) are providing insights into how the cell generates and maintains asymmetry.

5. *Secondary metabolism and development.* There is a considerable amount of descriptive information relating the secondary metabolism of plants and fungi to development. However, the role of alkaloids, pigments, and antibiotics remains an enigma. Recent advances in the cloning of *Streptomyces* genes for antibiotic synthesis suggests that a more penetrating understanding of their function and regulation may be forthcoming.

6. *Population differentiation.* It had once been a fond hope that heterocyst spacing in *Anabaena* would become the system for determining the mechanism of one-dimensional pattern formation. Apparently this was not to be. But all is not lost. The heterocystous cyanobacteria offer the clearest example among the prokaryotes of a population differentiated into two functionally distinct, coexisting, collaborating cell types (see Chapter 5). The opportunity to ask the obvious questions about regulatory switching, commitment, and differentiation with the background of understanding about promoter specificity and polymerase accessory factors (see Chapter 10) suggests that optimism about this problem is not an unreasonable attitude.

7. *Cell-to-cell interactions.* This aspect of cell biology is one of the roots of multicellularity and is fundamental to an understanding of embryogenesis. In addition, disruption of normal cell interactions is one of the proximal causes of malignant growth. Thus, it is no wonder that it is a subject of considerable interest and concern.[3] A great deal of information is available concerning interactions mediated by chemical messengers such as hormones and mating signals (see Chapter 11). Far less is known about the molecular mechanisms of contact-mediated interactions, such as those found in sea-urchin egg fertilization, sponge reaggregation, formation of the pseudoplasmodium of *Dictyostelium* and *Chlamydomonas* mating. Recently, there has been a burst of research using monoclonal antibodies against cell surface antigens to block cell interactions and subsequently to isolate and study the relevant signal molecules. Eventually, however, all of these systems stop short either because of the absence of a convenient system for genetic analysis or the difficulty in obtaining sufficient material to do biochemistry and molecular biology.

The area of cell interactions has traditionally been focused exclusively on eukaryotes, reflecting the fact that most of us have been taught that bacteria are unicellular creatures that do not interact or behave as social populations. While that seems to be true for *E. coli*, it certainly does not apply to the myxobacteria, the actinomycetes, the heterocystous cyanobacteria, and certainly many others yet to be studied. Over the past few years, the myxobacteria have emerged as the principal prokaryotic system for studying cell interactions,[4] and the existence of a variety of such interactions in the context of a system for convenient genetic and biochemical analyses offers the realistic hope of understanding such aspects as the nature of the signal-receptor interaction, regulation of signal system biosynthesis, and transduction of signal reception into a behavioral response.

In writing an epilogue such as this, it is tempting to close with a series of explicit summaries—to reduce the verbiage to a few quintessential conclusions. I have decided not to do so; for while such summaries are sometimes pedagogically useful, they imply that the contents of an entire field can be distilled into a few final bits worth remembering. Furthermore, such a summary would imply that there has emerged a program, a general theory, or a set of universals that describe development. Gunther Stent has made this point eloquently:

> . . . we cannot expect to discover a general theory . . . of development, no more than historians any longer expect to discover, as some once did, a general theory of history. Rather, we are faced with a near infinitude of particulars which have to be sorted out, case by case.

> ... right now, it seems to be a good time for molecular biologists to iron out some very important details, by putting some flesh and bones into the abstract theories of the developmental systems analysts. These details may not lead us to biological universals comparable in standing to the central dogma or the genetic code, but they will illuminate multi-dimensional historical phenomena that are much more complex, and possibly even more fascinating, than the one-dimensional, programmatic phenomenon of protein synthesis.[5]

I think of the work described in this book, and of the body of science in general, not as a process that leads inexorably to a series of final conclusions but rather one that goes on—perhaps never ending—occasionally providing an insight, a generalization, occasionally misleading, but generally providing more light than darkness. But it is the entire process that is the final product—incomplete and imperfect but always getting better.

References

Preface

1. An indication of the fact that microbial development represents a bona fide (and fast-moving) aspect of developmental biology is a symposium recently held at Cold Spring Harbor (May 11-15, 1983). In this symposium, "Microbial Development," 70 papers were presented on such systems as *Bacillus* sporulation, cellular differentiation in *Caulobacter*, cell interactions in myxobacteria, chemotaxis, cell division in *E. coli*, *Streptomyces* development, yeast mating, *Dictyostelium* development, as well as a number of other systems. An outcome of this meeting was a compilation of review articles by some of the participants, published as Microbial development (R. Losick and L. Shapiro, eds.). Cold Spring Harbor Laboratory, N.Y., 1984.

Chapter 1 Introduction

1. Jacob's quote is translated from his book, *La logique du vivant: Une histore de l'hérédité*, (Gallimard, Paris, 1970).
2. See an essay on the subjects of pleomorphism, monomorphism, Robert Koch, Sergei Winogradsky, and their roles in shaping the modern view of microbiology; Penn, M. and Dworkin, M. 1976. Robert Koch and two visions of microbiology. Bact. Rev. 40:276-83.
3. Jacob, F. and Monod, J. 1963. Genetic repression, allosteric inhibition and cellular differentiation. *In* Cytodifferentiation and macromolecular synthesis (M. Locke, ed.). Academic Press, New York, London, p. 30.

4. Lederberg, J. 1966. *In* Current topics in developmental biology, Vol. 1, (A. A. Moscona and A. A. Monroy, eds.). Academic press, New York, London, p. ix. In a brief and provocative essay, Gunther Stent has argued that simply understanding the differential expression of developmental genes is far from an understanding of the three-dimensional fabric of a developing cell. "The noxious impact of molecular biology came about because the tenet that the gene is a one-dimensional description of the primary structure of a particular protein molecule was turned, willy-nilly, into the doctrine that the genome is a one-dimensional description of the whole animal. In particular, it came to be believed that the genome embodies, not merely a protein catalog, but a genetic program for development, from zygote to adult." Stent, G. 1985. Thinking in one dimension: the impact of molecular biology on development. Cell, *40*:1-2.
5. From a lecture by H. Echols.
6. See Whittenbury, R. and Dow, C. S. 1977. Morphogenesis in bacteria. *In* Companion to microbiology: Selected papers for further study (A. T. Bull and P. M. Meadow, eds.). Tongman, Inc., N.Y., pp. 221-63.
7. The late T. M. Sonneborn has written a thoughtful essay on development: Cell differentiation and communication: Patterns, problems, and probes. *In* Microbiology 1975, (D. Schlessinger, ed. 1975). ASM, Washington, D.C., pp. 421-25.

Some general and useful review and background reading:

(A) Bonner, J. T. 1974. On development: The biology of form. Harvard University Press, Cambridge. In this book, Professor Bonner, one of the real scholars in our field, examines the role of evolution in development, wonders how temporal and spatial aspects of development can be encoded in DNA, and generally tries to fold molecular biology into the larger developmental questions.
(B) Parish, J. H. (eds.). 1979. Developmental biology of prokaryotes. Univ. of California Press, Berkeley and Los Angeles. This book contains a collection of chapters on the developmental biology of the bacteria.
(C) Dworkin, M. 1979. Spores, cysts and stalks. *In* The bacteria, Vol. 7 (J. R. Sokatch and L. N. Ornston, eds.). Academic press, pp. 2-84. This work reviews spores and cyst morphogenesis in *Bacillus*, Actinomycetes, myxobacteria, *Azotobacter*, and cyanobacteria; it also presents a discussion of *Caulobacter* stalk morphogenesis.
(D) Saier, M. H. and Jacobson, G. R. 1984. The molecular basis of sex and differentiation. Springer-Verlag, N.Y. This is an excellent and thoughtful book that treats sex, differentiation, and death as developmental processes and examines them in the context of a variety of prokaryotic and eukaryotic microbes.

Chapter 2 Regulation and Morphogenesis in Bacteriophage Development

1. Kruger, D. H. and Schroeder, C. 1981. Bacterophage T3 and bacteriophage T7 virus-host cell interactions. Microbiol. Rev. *45*:9-51.

2. See a series of four articles in *Microbiology 1983* (D. Schlessinger, ed.). ASM, Washington, D.C. These are under the general heading "Antitermination: The lambda system" and are included in pp. 35–52. Then, of course, there is the lambda bible, the proceedings of an historic conference at Cold Spring Harbor in 1970: The Bacteriophage lambda (A. D. Hershey, ed.). Cold Spring Harbor Laboratory, 1971.
3. Losick, R. and Pero, J. 1976. Regulatory subunits of RNA polymerase. *In* RNA polymerase. Cold Spring Harbor Laboratory, N.Y., pp. 227–46.
4. See ref. 3. See also Koerner, J. F. and Snustad, D. P. 1979. Shutoff of host macromolecular synthesis after T-even bacteriophage infection. Microbiol. Rev. 43:199–223.
5. In the introduction to their review on bacteriophage (Herskowitz, I. and Hagen, D. 1980). The lysis-lysogeny decision of phage: Explicit programming and responsiveness. Ann Rev. Genet. 14:399–445), the authors state, "We recently spoke with someone who told us that he would sooner read a paper on algebraic topology than on λ." The authors have written an exception to the rule—that is, a highly readable review on λ.
6. There is a brief review of some aspects of phage Mu (1983. Transposable genetic elements: The bacteriophage Mu paradigm. ASM News, 49:#6: 275–80) and a more comprehensive review (1976. Bacteriophage Mu as a transposition element. Ann. Rev. Genet. 10:389–411) both by A. I. Bukhari.
7. Silvermann, M. and Simon, M. 1980. Phase variation: Genetic analysis of switching mutants. Cell 19:845–54.
8. Giphart-Gassler, M., Plasterk, R.H.A., and Van de Putte, P. 1982. G inversion in bacteriophage Mu: A novel way of gene splicing. Nature 279:339–42. Also see Howe, M. M. 1980. The invertible G segment of phage Mu. Cell 21:605–6 (This is a "mini-review" occasionally published by Cell).
9. Murialdo, H. and Becker, A. 1978. Head morphogenesis of complex double-stranded deoxyribonucleic acid bacteriophages. Microbiol. Rev. 42:529–76.

Chapter 3 The Endospore

1. The two papers describing the discovery of endospores in *Bacillus* and their role in the life cycle are:
   (A) Cohn, F. 1876. Untersuchungen über Bakterien: IV Beiträge zur Biologie der Bacillen. Beiträge zur Biologie der Pflanzen. Vol. 2:249–76.
   (B) Koch, R. 1876. Die Aetiologie der Milzbrand-Krankheit, begründet auf die Entwicklungsgeschichte des *Bacillus anthracis*. Beiträge zur Biologie der Pflanzen, Vol. 2, No. 2;277–310.
   Translation of both of these papers, as well as a number of other important historical papers in microbiology, can be found in Brock, T. 1961. Milestones in microbiology. Prentice-Hall, Englewood Cliffs, N.J.
2. Review articles that generally cover the area of spore structure are:
   (A) Warth, A. D. 1978. Molecular structure of the bacterial spore. Adv. Microb. Physiol. 17:1–45. If you are going to read just one detailed and comprehensive review, this is the one.

(B) Tipper, D. J. and Gauthier, J. J. 1972. Structure of the bacterial endospore. *In* Spores V (H. R. Halvorson, R. Hanson, and L. L. Campbell, eds.). ASM, Washington, D.C., pp. 3–12.

(C) Dworkin, M. 1979. Spores, cysts and stalks. *In* The bacteria, Vol. 7 (J. R. Sokatch and L. N. Ornston, eds), pp. 1–84.

In addition, some specific references will be presented for some aspects of spore structure.

3. Henner, D. S. and Hoch, H. A. 1980. The *Bacillus subtilis* chromosome. Microbiol. Rev. *44*:57–82.

4. An excellent, brief summary of the question of the size of the genome of the *Bacillus* spore is found in Wake, R. G., 1980. How many chromosomes in the *Bacillus subtilis* spore and of what size? Spore Newsletter, VII, No. 2:21–26.

5. Murrell, W. G and Warth, A. D. 1964. Composition and heat resistance of bacterial spores. *In* Spores III (L. L. Campbell and H. O. Halvorson, eds.). ASM, Washington, D.C., pp. 1–24. The Spore Conferences are held every three years in the U.S. (the last one was held in 1984); the proceedings are published as a series of brief articles or reviews and are an excellent reflection of the current work on endospores.

6. Setlow, P. 1981. Biochemistry of bacterial forespore development and spore germination. *In* Sporulation and germination (H. Levinson, A. L. Sonenshein, and D. J. Tipper, eds.). ASM, Washington, D.C., pp. 13–28. this work is a general review of the metabolic events occurring during germination, including a discussion of the germination proteins and their proteases.

7. Aronson, A. I. and Fitz-James, P. C. 1976. Structure and morphogenesis of the bacterial spore coat. Bact. Rev. *40*:360–402.

8. Wilkinson, B. J., Deans, J. A., and Ellar, D. J. 1975. Biochemical evidence for the reversed polarity of the outer membrane of the bacterial forespore. Biochem. J. *152*:561–69.

9. Three additional review articles on germination are:

   (A) Dawes, I. W. and Hansen, J. N. 1972. Morphogenesis in sporulating bacilli. Crit. Rev. Microbiol. *1*:479–520.

   (B) Gould, G. W. and Dring, G. J. 1972. Biochemical mechanisms of spore germination. *In* Spores V (H. O. Halvorson, R. Hanson, and L. L. Campbell, eds.). ASM, Washington, D.C., pp. 401–8.

   (C) Setlow, P. 1983. Germination and outgrowth. *In* The bacterial spore, vol. 2 (A. Hurst and G. W. Gould, eds.). Academic Press, London, pp. 211–244.

10. For this discussion I have drawn heavily on three excellent reviews:

    (A) Piggot, P. J. and Coote, D. 1976. Genetic aspects of bacterial endospore formation. Bact. Rev. *40*:908–62.

    (B) Piggot, P. J. 1979. Genetic strategies for studying bacterial differentiation. Biol. Rev. *54*:347–67.

    (C) Piggot, P. J., Moir, A, and Smith, D. A. 1981. Advances in the genetics of *Bacillus subtilis* differentiation. *In* Sporulation and germination (H. Levinson, A. L. Sonenshein, and D. J. Tipper, eds.). ASM, Washington, D.C., pp. 29–39.

10A. The recent paper by Piret, J. M. and Demain, A. L. 1983. (J. Gen. Microbiol. *129*:1309–16) challenges the reported relationship between gramicidin S and

sporulation and spore properties. It also contains in its list of references citations from Henry Paulus's laboratory and others that originally suggested the relationship.
11. Lancastre, H. and Piggot, P. J. 1979. Identification of different sites of expression for *spo* loci by transformation of *Bacillus subtilis*. J. Gen. Microbiol. *114*:377–89.
12. Dancer, B. N. 1981. Control of sporulation in fused protoplasts of *Bacillus subtilis*, 168. J. Gen. Microbiol. *126*:29–36.
13. Lovett, P. S. 1981. Cloning strategies in *Bacillus subtilis*. *In* Sporulation and germination (H. Levinson, A. L. Sonenshein, and D. J. Tipper, eds.). ASM, Washington, D.C., pp. 40–47.
14. (A) Hemphill, H. E. and Whiteley, H. R. 1975. Bacteriophages of *Bacillus subtilis*. Bacteriol. Rev. *38*:257–315.
   (B) Reilly, B. E. 1976. Bacteriophage of *Bacillus subtilis* as a paradigm of bacteriophage of Bacilli. *In* Microbiology 1976 (D. Schlessinger, ed.). ASM, Washington, D.C., pp. 228–37.
15. For a description of SPβ and its use as a specialized transducing phage, see Zahler, S. 1982. Specialized transduction in *Bacillus subtilis*. *In* The molecular biology of bacilli (D. Dubnau, ed.). Academic Press, New York, p. 269. For a description of the use of the transposon Tn917, see Youngman, P. Y., Perkins, J. B., and R. Losick. Genetic transpotition and insertional mutagenesis in *Bacillus subtilis* with the *Streptococcus faecalis* transposon Tn917. Proc. Nat. Acad. Sci., U.S.A. *80*:2305–9, 1983.
16. This phenomenon, referred to as "endotrophic sporulation," was described by J. W. Foster and J. J. Perry (1954. Intracellular events occurring during endotrophic sporulation in *Bacillus mycoides*. J. Bacteriol. *67*:295–302) and represented the beginning of modern studies on the physiology of initiation of sporulation.
17. Schaeffer, P., Millet, J. and Aubert, J-P. 1965. Catabolic repression of bacterial sporulation. Proc. Natl. Acad. Sci., U.S.A. *54*704–11. Ideas about catabolite repression in *E. coli* were surging forward, and the analogy with *Bacillus* sporulation was a tempting one. It turned out to be an inappropriate analogy, but like so many other intelligent wrong ideas, it led off in other useful directions.
18. Mandelstam, J. and Higgs, S. A. 1974. Induction of sporulation during synchronized chromosome replication in *Bacillus subtilis*. J. Bacteriol. *120*:38–42.
19. One of the earliest of these was initially referred to as "endogenous factor" (Srinivasan, V. R. and Halvorson, H. O. 1963. Endogenous factor in sporogenesis in bacteria. Nature *197*:100–1) and later as *sporogen* (Srinivasan, V. R. 1966. Sporogen: An inductor for bacterial cell differentiation. Nature *209*:537).
20. Freese, E. and Frujita, Y. 1976. Control of enzyme synthesis during growth and sporulation. *In* Microbiology 1976 (D. Schlessinger, ed.). ASM, Washington, D.C., pp. 164–84.
21. See the chapter by G. H. Chambliss (1979. The molecular biology of *Bacillus subtilis* sporulation. *In* Developmental biology of prokaryotes [J. H. Parish, ed.]. University of California Press, Berekely and Los Angeles, pp. 57–71) for a nice discussion of the various regulatory mechanisms that might be involved in the control of *Bacillus* sporulation.

22. Freese, E. and Heinze, J. 1983. Metabolic and genetic control of bacterial sporulation. *In* The bacterial spore, Vol. 2 (A. Hurst and G. W. Gould, eds.). Academic Press, p. 101-72. This is an excellent and up-to-date review that relates the biochemistry and the morphology of sporulation.
23. Hardwick, W. A. and Foster, J. W. 1952. On the nature of sporogenesis in some aerobic bacteria. J. Gen. Physiol. *35*:907-27.
24. Sterlini, J. M. and Mandelstam, J. 1969. Commitment to sporulation in *Bacillus subtilis* and its relationship to development of actinomycin resistance. Biochem. J. *113*:29-37.
25. Chambliss (see ref. 21) presents the evidence that sporulation involves the regulated expression of sporulation genes.
26. In a small and provocative book, B. Wright challenges the notion that all that is necessary for development to proceed is to regulate the timed and orderly expression of developmental genes. See *Critical variables in differentiation,* Prentice-Hall, N.J., 1973.
27. Hanson, R. S., Peterson, J. A., and Yousten, A. A. 1970. Unique biochemical events in bacterial sporulation. Ann Rev. Microbiol. *24*:53-90.
28. Losick, R. and Sonenshein, A. L. 1969. Change in the template specificity of RNA polymerase during sporulation of *Bacillus subtilis.* Nature *224*:35-37.
29. See the brief review by R. Losick for a quick overview of this elegant and very important approach to understanding the regulation of sporulation. (1981. Sigma factors, stage 0 genes and sporulation. *In* Sporulation and germination [H. Levinson, A. L. Sonenshein, and D. J. Tipper, eds.] ASM, Washington, D.C., pp. 48-56.) Also, strongly recommended is the recent review by Losick, R. and Youngman, P. 1984. Endospore formation in *Bacillus. In* Microbial development (R. Losick and L. Shapiro, eds.). Cold Spring Harbor, pp. 63-88.
30. Doi, R. H. 1977. Genetic control of sporulation. Ann. Rev. Genetics *11*:29-48.
31. Leighton, T. J. 1973. An RNA polymerase mutation causing temperature-sensitive sporulation in *Bacillus subtilis.* Proc. Nat. Acad. Sci., U.S.A. *70*:1179-83.
32. See, for example, Orrego, C., Kerjan, P., Manca de Nadra, M. C., and Szulmajster, J. 1973. Ribonucleic acid polymerase in a thermosensitive sporulation mutant (ts-4) of *Bacillus subtilis.* J. Bacteriol. *116*:636-47.
33. Tipper, D. J., Johnson, C. W., Ginther, C. L., Leighton, T., and Wittman, H. G. 1977. Erythromycin resistant mutations in *Bacillus subtilis* cause temperature sensitive mutations. Molec. Gen. Genet. *150*:147-59.
34. W. G. Murrell has written a concise but comprehensive review of the subject of dormancy and resistance: 1981. Biophysical studies on the molecular mechanisms of spore heat resistance and dormancy. *In* Sporulation and germination (H. Levinson, A. L. Sonenshein, and D. J. Tipper, eds.). ASM, Washington, D.C., pp. 64-77. For a more extensive coverage of this subject see the Spore Newsletter (Vol. VII, #5, May 1981), which contains a collection of papers presented at a recent workshop on the subject "Bases and Mechanisms of Bacterial Spore Resistance." Perhaps the most detailed and encyclopedic coverage of the subject of dormancy and resistance can be found in the chapter by Gould, G. W. 1983. Mechanisms of resistance and dormancy. *In* The bacterial spore, vol. 2 (A. Hurst and G. W. Gould, eds.). Academic Press, London, pp. 211-254.

Chapter 4   Caulobacter

1. Henrici, A. T. and Johnson, D. T. 1935. Studies of freshwater bacteria: II. Stalked bacteria, a new order of Schizomycetes. J. Bacteriol. 30:61-93.
2. Poindexter, J. S. 1964. Biological properties and classification of the *Caulobacter* group. Bacteriol. Rev. 28:231-95.
3. Two recent reviews on *Caulobacter* are:
   (A) Ely B. and Shapiro, L. 1984. Regulation of cell differentiation in *Caulobacter crescentus*. In Microbial development (R. Losick and L. Shapiro, eds.). Cold Spring Harbor, pp. 1-26.
   (B) Poindexter, J. S. 1981. The Caulobacters: Ubiquitous unusual bacteria. Microbiol. Rev. 45:123-79.
   The Ely and Shapiro review focuses on development; the Poindexter review is far more extensive and detailed and covers all aspects of the biology of *Caulobacter*, including development.
4. Dow, C. S. and Whittenbury, R. 1983. The cell cycle of *Rhodomicrobium vannielii*: Obligate morphogenesis and differentiation. In Microbiology 1983 (D. Schlessinger, ed.). ASM, Washington, D.C., pp. 167-69. This is a very brief overview of one aspect of the development of *R. vannielii*. For a more detailed account see: Whittenbury, R. and Dow, C. S. 1977. Morphogenesis and differentiation in *Rhodomicrobium vannielii* and other budding and prosthecate bacteria. Bacteriol. Rev. 41:754-808.
5. Newton, A. 1984. Temporal and spatial control of the *Caulobacter* cell cycle. In The microbial cell cycle (P. Nurse and E. Streiblora, eds.). CRC Press, Boca Raton, Fla. This is an excellent and very thoughtful discussion of *Caulobacter* development.
6. Donachie, W. D. 1979. The cell cycle of *Escherichia coli*. In Developmental biology of prokaryotes. University of California, Berkeley and Los Angeles, pp. 11-35. Discussions of prokaryote development often include a section on the regulation of growth and division in *E. coli*. I have chosen not to do so for reasons briefly discussed in the Introduction. However, the issue of control of cell division is of particular relevance to development in *Caulobacter*, and the reader is directed to Donachie's review.
7. Ely has presented evidence that indicates that there are, in fact, three flagellins produced by *C. crescentus* (Johnson, R. C., Ferber, D. M., and Ely, B. 1983. Synthesis and assembly of flagellar components by *Caulobacter crescentus* motility mutants. J. Bacteriol. 154:1137-44).
8. Shapiro, L. and Bendis, I. 1975. RNA phages in bacteria other than *E. coli*. In RNA phages (N. Zinder, ed.). Cold Spring Harbor Monograph Series, Cold Spring Harbor Laboratory, N.Y., pp. 397-410.
9. Sollick, J. D. 1972. Differential phage sensitivity of cell types in *Caulobacter*. J. Gen. Virol. 16:405-7.
10. Shapiro, L., Nisen, P., and Ely, B. 1981. Genetic analysis of the differentiating bacterium, *Caulobacter crescentus*. In Society of General Microbiology Symposium 31:317-39. This is the most recent (and in fact, the only) review of the genetics of *Caulobacter*. For a more recent and detailed description of mapping and the genetic map in *Caulobacter*, see:

(A) Barrett, J. T., Croft, R. H., Ferber, D. M., Gerardot, C. J., Schoenlein, P. V., and Ely, B. 1982. Genetic mapping with Tn5-derived auxotrophs of *Caulobacter crescentus*. J. Bacteriol. *151*:888–98.

(B) Barrett, J. T., Rhodes, C. S., Ferber, D. M., Jenkins, B., Kuhl, S. A., and Ely, B. 1982. Construction of a genetic map for *Caulobacter crescentus*. Ibid., *149*:889–96.

11. Gomes, S. L. and Shapiro, L. 1984. Differential expression and positioning of chemotaxis methylation proteins in *Caulobacter*. J. Mol. Bol. In press.
12. Milhausen, M. and Agabian, N. 1983. *Caulobacter* Flagellin mRNA segregates asymmetrically at cell division. Nature *302*:630–32.

Chapter 5 Heterocyst Development in Cyanobacteria

1. For a detailed and comprehensive account of the cyanobacteria, see the recent book, *The biology of cyanobacteria* (1982. [N. G. Carr and B. A. Whitton, eds.]. Univ. of California Press, Berkeley and Los Angeles.) While it does not emphasize the developmental aspects of the cyanobacteria, the book contains several chapters that do.
2. Stanier, R. Y. and Cohen-Bazire, G. 1977. Phototrophic prokaryotes: The cyanobacteria. Ann. Rev. Microbiol. *31*:225–74. If you have the time to read only one background review on the biology of the cyanobacteria, this is probably the one.
3. Two excellent reviews focusing on the physiology, biochemistry, and development of heterocysts are:

    (A) Haselkorn, R. 1978. Heterocysts. Ann Rev. Plant Physiology *29*:319–44.

    (B) Wolk, C. P. 1982. Heterocysts. *In* The biology of cyanobacteria (N. G. Carr and B. A. Whitton, eds.). Univ. of California Press, Berkeley and Los Angeles, pp. 359–86.

4. Thomas, J., Meeks, J. C., Wolk, C. P., Shaffer, P. W., Austin, S. M., and Chien, W. S. 1977. Formation of glutamine from [^{13}N] ammonia, [^{13}N] dinitrogen, and [^{14}C] glutamate by heterocysts isolated from *Anabaena cylindrica*. J. Bacteriol. *129*:1545–55. This is an important paper that not only described a method for isolating relatively pure suspensions of heterocysts that were still enzymatically active but also worked out a number of apsects of the flow of carbon and nitrogen between heterocysts and vegetative cells.
5. Simon, R. D. 1980. DNA content of heterocysts and spores of the filamentous cyanobacterium *Anabaena variabilis*. F.E.M.S. Microbiol. Lett. *8*:241–45.
6. Fay, P., Stewart, W.D.P., Walsby, A. E., and Fogg, G. E. 1968. Is the heterocyst the site of nitrogen fixation in blue-green algae? Nature *220*:810–12.
7. Peterson, R. B. and Wolk, C. P. 1978. High recovery of nitrogenase activity and of ^{55}Fe-labeled nitrogenase in heterocysts isolated from *Anabaena variabilis*. Proc. Nat. Acad. Sci., U.S.A. *75*:6271–75.
8. Haury, J. F. and Wolk, C. P. 1978. Classes of *Anabaena variabilis* mutants with oxygen-sensitive nitrogenase activity. J. Bacteriol. *136*:688–92.
9. Wolk, C. P. 1968. Movement of carbon from vegetative cells to heterocysts in *Anabaena cylindrica*. J. Bacteriol. *96*:2138–43.

10. Bradley, S. and Carr, N. G. 1976. Heterocyst and nitrogenase development in *Anabaena cylindrica.* J. Gen. Microbiol. *96*:175-84.
11. Fogg, G. E. 1949. Growth and heterocyst production in *Anabaena cylindrica,* Lemm. II. In relation to carbon and nitrogen metabolism. Ann. Botan, N.S. *13*:241-59.
12. Fisher, R., Tuli, R., and Haselkorn, R. 1981. A cloned cyanobacterial gene for glutamine synthetase functions in *E. coli,* but the enzyme is not adenylated. Proc. Nat. Acad. Sci., U.S.A. *78*:3393-97.
13. (A) Wilcox, M. 1970. One-dimensional pattern found in blue-green algae. Nature *228*:686-87.
    (B) Wilcox, M., Mitchison, G. J., and Smith, R. J. 1973. Pattern formation in the blue-green alga *Anabaena:* I. Basic mechanisms. J. Cell Sci. *12*:707-23.
    (C) Wilcox, M., Mitchison, G. J., and Smith, R. J. 1973. Pattern formation in the blue-green alga *Anabaena:* II. Controlled proheterocyst regression. J. Cell Sci. *13*:637-49.
    (D) Wilcox, M., Mitcheson, G. J., and Smith, R. J. 1975. Spatial control of differentiation in the blue-green alga *Anabaena. In* Microbiology 1975 (D. Schlessinger, ed.). ASM, Washington, D.C., pp. 453-63.
    (E) Wilcox, M., Mitcheson, G. J., and Smith, R. J. 1975. Mutants of *Anabaena cylindrica* altered in heterocyst spacing. Arch. Microbiol. *103*:219-23.
    This remarkable series of papers illustrates the power of carefully observing cells and doing a few simple, clever experiments.
14. See ref. 3 (B) for the arguments for stimulation vs. inhibitor.
15. Wolk, C. P., Vonshak, A., Kehoe, P., and Elkai, P. 1984. Construction of shuttle vectors capable of conjugative transfer from *Escherichia coli* to nitrogen-fixing filamentous cyanobacteria. Proc. Nat. Acad. Sci., U.S.A., in press.
16. Turner, N. E., Robinson, S. J., and Haselkorn, R. 1983. Different promoters for the *Anabaena* glutamine synthetase gene during growth using molecular or fixed nitrogen. Nature *306*:337-42.

Chapter 6 Streptomyces

1. Kalakoutskii, L. V. and Agre, N. S. 1976. Comparative aspects of development and differentiation in Actinomycetes. Bacteriol. Rev. *40*:469-524.
2. Kendrick, K. E. and Ensign, J. 1983. Sporulation of *Streptomyces griseus* in submerged culture. J. Bacteriol. *155*:357-66.
3. Two excellent reviews on the actinomycetes are:
   (A) Ensign, J. C. 1978. Formation, properties and germination of actinomycete spores. Ann. Rev. Microbiol. *32*:185-219. This review emphasizes the physiology of spore development and includes *Streptomyces, Thermoactinomyces,* and the Actinoplanaceae.
   (B) Chater, K. F. and Merrick, M. J. 1979. Streptomyces. *In* Developmental biology of prokaryotes (J. H. Parish, ed.). Univ. of California Press, Berkeley and Los Angeles, pp. 93-114. This review focuses on the genus *Streptomyces* and emphasizes the genetics of *S. coelicolor.*

4. Hirsch, C. F. and Ensign, J. C. 1976. Nutritionally defined conditions for germination of *Streptomyces viridochromogenes* spores. J. Bacteriol. 126:13–23.
5. There are two reviews on *Streptomyces* antibiotics that cover two different aspects of the problem:
   (A) Martin, J. F. and Demain, A. L. 1980. Control of antibiotic synthesis. Microbiol. Rev. 44:230–251. This review focuses mainly on the antibiotics of *Streptomyces* but is concerned with the general problems of physiology and regulation of antibiotic synthesis.
   (B) Gottlich, S. 1976. The production and role of antibiotics in soil. J. Antibiotics, Tokyo, 29:987–1000. This review presents the ecological arguments concerning the natural role of antibiotics.
6. For example, McCann, P.A. and Pogell, B. M. (1979. Panamycin: A new antibiotic and stimulator of antibiotic formation. J. antibiotics 32:673–78) have shown that an endogenously generated antibiotic seems to play a regulatory role in *Streptomyces* development. See J. C. Ensign (1976. Properties and germination of Streptomyces spores and a suggestion for the function of antibiotics. *In* Microbiology 1976 [D. Schlessinger, ed.], ASM Press, Washington, D.C., pp. 531–33) for a discussion of the possible role of some *Streptomyces* antibiotics as germination regulators.
7. This discussion is based mainly on the review by K. F. Chater and M. J. Merrick referred to in ref. 3(B).
   The reader is also directed to two recent reviews from Chater's laboratory:
   (A) Lomovskaya, N. D., Chater, K. F., and Mkrtumian, N. M. 1980. Genetics and molecular biology of *Streptomyces* bacteriophages. Microbiol. Rev. 44:206–29. The authors open their review with the upbeat statement, "The streptomyces are genetically among the most tractable of bacteria." This claim is based, in part, on the fact that there is available a *Streptomyces* phage, ϕC31, which is a temperate phage that has been useful for the genetic analysis of *Streptomyces*. The authors also predict that the phage will be useful in searching for and transferring transposons as well as a vector for introducing foreign DNA into *Streptomyces* hosts.
   (B) Chater, K. 1984. Morphological and physiological differentiation in *Streptomyces*. *In* Microbial development (R. Losick and L. Shapiro, eds.). Cold Spring Harbor, N.Y.
8. See J. C. Ensign's review referred to in ref. 3(A) for an overview of metabolic and biosynthetic events occurring during spore germination.
9. Mikulik, K., Jauda, I., Ricicova, A., and Vinter, A. 1975. Differentiation in actinomycetes: I. Ribosomes of vegetative cells and spores of *Streptomyces granaticolor*. *In* Spores VI (P. Gerhardt, R. N. Costilow, and H. L. Sadoff, eds.). ASM, Washington, D.C., pp. 15–27.
10. Sonenshein, A. L. and Losick, R. 1970. RNA polymerase mutants blocked in sporulation. Nature 227:906–9.
11. Chater, K. F. 1974. Rifampicin-resistant mutants of *Streptomyces coelicolor* A3(2). J. Gen. Microbiol. 80:277–90. These results have recently been confirmed in *S. griseus* by J. C. Ensign (see ref. 2) who showed that among a number of mutants resistant to thiostrepton, rifampicin, or neomycin, none showed any defect in sporulation.
12. Autsnov, P. P., Ivanov, I. G. and Markov, G. G. 1977. Heterogeneity of streptomyces DNA. F.E.B.S. Lett. 79:151–54.

13. Khokhlov, A. S. and Tovarova, I. I. 1979. Autoregulator from *Streptomyces griseus*. F.E.B.S. Lett. *55*:133-45. The work on cloning the A factor gene(s) is reported by Horinouchi, S., Kumoda, Y., and Beppu, T. (1984). Unstable genetic determinant of A-factor biosynthesis in streptomycin-producing organisms: Cloning and characterization. J. Bacteriol. 158:481-87.
14. Bero, S., Bekesi, I., Vitalis, S., and Szabo, G. 1980. A substance effecting differentiation in *Streptomyces griseus*: purification and properties. Eur. J. Bioch. *103*:359-64.
15. Stanley Cohen at Stanford and Richard Losick at Harvard are both looking at various molecular aspects of development in *Streptomyces*. For example, Westpheling in Losick's laboratory has recently shown that the RNA polymerase of *S. coelicolor* can recognize developmental promoter sequences in cloned developmental genes of *Bacillus subtilis* and transcribe those genes. These interesting results were presented at the recent Cold spring Harbor symposium on "Microbial Development" (May 11-15, 1983). Also see reference 14, Chapter 10.

Chapter 7 The Myxobacteria

1. Thaxter, R. 1892. On the Myxobacteriaceae, a new order of Schizomycetes. Botan., Gazette *17*:389-406.
2. Thaxter, R. 1904. Notes on the Myxobacteriaceae. Botan. Gazette *37*:405-16.
3. Jahn, E. 1924. Beiträge zur botanischen Protistologie: I. Die Polyangiden. Gebruder Borntrager, Leipzig.
4. Woods, N. A. 1948. Studies on the Myxobacteria. Master's thesis, Univ. of Washington, Seattle.
5. There are a series of relatively recent reviews on the myxobacteria that emphasize different aspects of their biology and development.
    (A) Kaiser, D., Manoil, C., and Dworkin, M. 1979. Myxobacteria: Cell interactions, genetics and development. Ann. Rev. Microbiol. *33*:595-739. This review generally covers the area of myxobacterial development.
    (B) Dworkin, M. and Kaiser, D. 1985. Cell interactions in myxobacterial growth and development. Science, in press. This review focuses on the various cell interactions that are characteristic of the myxobacteria.
    (C) Zusman, D. R. 1980. Genetic approaches to the study of development in the myxobacteria. *In* The molecular genetics of development (T. Leighton and W. F. Loomis, eds.). Academic Press, N.Y., pp. 41-78. This is a detailed review of the genetics of myxobacteria. It should be emphasized, however, that this area is so fast-moving that in the four years that intervened between the above review and this book a number of important genetic techniques and analyses of myxobacterial development have emerged. For the most recent description of myxobacterial genetics, see the chapter by D. Kaiser (Genetics of myxobacteria, pp. 163-184) in ref. 5(F).
    (D) White, D. 1981. Cell interactions and the control of development in myxobacteria populations. Int'l Rev. Cytol. *71*:203-27. This review is also of a general nature but includes somewhat more of the work on *Stigmatella*.

(E) Reichenbach, H. and Dworkin, M. 1981. The order Myxobacterales. *In* The prokaryotes: A handbook on habitats, isolation and identification of bacteria (M. P. Starr, H. Stolp, H. G. Truper, A. Balows, and H. G. Schlegel, eds.). Springer-Verlag, Berlin, pp. 328–55. This review is concerned with the more biological aspects of the myxobacteria as a general group. It covers ecology, taxonomy, isolation, cultivation, etc.

(F) Rosenberg, E. (ed.) 1984. Myxobacteria: development and cell interactions. Springer-Verlag, N.Y. This book is an outgrowth of the Tenth annual meeting on the "Biology of the Myxobacteria" (Jerusalem, 1983) and is certainly the most comprehensive and up-to-date compilation and review of work on the myxobacteria.

6. See ref. 5(E).
7. Rosenberg, E., Keller, K. H., and Dworkin, M. 1977. Cell density-dependent growth of *Myxococcus xanthus* on casein. J. Bacteriol. *129*:770–77.
8. See ref. 2. Also see reference 5(E) for the taxonomic scheme referred to here. An alternative scheme has been proposed by McCurdy, H. (1974. Myxobacterales. *In* Bergey's manual of determinative bacteriology, 8th ed. [R. E. Buchanan and N. E. Gibbons, eds.]. Williams & Wilkins, Baltimore, M.D., pp. 76–98).
9. Orndorff, P. E. and Dworkin, M. 1980. Separation and properties of the cytoplasmic and outer membranes of vegetative cells of *Myxococcus xanthus*. J. Bacteriol. 149:29–39.
10. White, D., Dworkin, M., and Tipper, D. J. 1968. Peptidoglycan of *Myxococcus xanthus*: Structure and relation to morphogenesis. J. Bacteriol. *95*:2186–97.
11. MacRae, T. H., Dobson, W. J., and McCurdy, H. D. 1977. Fimbriation in gliding bacteria. Canad. J. Microbiol. *23*:1096–1108.
12. See ref. 10.
13. Dworkin, M. and Gibson, S. M. 1964. A system for studying microbial morphogenesis: Rapid formation of microcysts in *Myxococcus xanthus*. Science *146*:243–44.
14. Kottel, R. H., Bacon, K., Clutter, D., and White, D. 1975. Coats from *Myxococcus xanthus*: Characterization and synthesis during myxospore differentiation. J. Bacteriol. *124*:550–57.
15. (A) Inouye, S., Ike, Y., and Inouye, M. 1983. Tandem repeat of the genes for protein S, a development specific protein of *Myxococcus xanthus*. J. Biol. Chem. *258*:38–40.

    (B) Downard, J. S., Kupfer, D., and Zusman, D. 1984. Gene expression during development of *Myxococcus xanthus*: Analysis of the genes for protein S. J. Mol. Biol. *175*:469–92.
15A. Sutherland, I. W. and Thomson, S. 1975. Comparison of polysaccharide produced by *Myxococcus* strains. J. Gen. Microbiol. *89*:124–32.
16. Sutherland, I. W. and Mackenzie, C. L. 1977. Glucan common to the microcyst walls of cyst-forming bacteria. J. Bacteriol. *129*:599–605.
17. See ref. 14.
18. The reader who is interested in the function of surface polysaccharides in bacteria is directed to an excellent review on the subject of W. F. Dudman (1977. The role of surface polysaccharides in natural environments. *In* Surface carbohydrates of the prokaryotic cell [I. W. Sutherland, ed.]. Academic Press, pp. 357–414).

19. D. A. Rees has written a very provocative essay on the subject of self-assembly of mixtures of polysaccharides: 1972. Shapely polysaccharides. Bioch. J. 126:257–73.
20. Burchard, R. P. 1981. Gliding motility of prokaryotes: Ultrastructure, physiology and genetics. Ann. Rev. Microbiol. 35:497–529.
21. (A) Keller, K. H., Grady, M., and Dworkin, M. 1983. Surface tension gradients: Feasible model for gliding motiity of Myxococcus xanthus. J. Bacteriol. 155:1358–66.
    (B) Dworkin, M., Weisberg, D., and Keller, K. H. 1983. Experimental observations consistent with a surface tension model of gliding motility of Myxococcus xanthus. J. Bacteriol. 155:1367–71.
22. Hodgkin, J. and Kaiser, D. 1979. Genetics of gliding motility in Myxococcus xanthus (Myxobacterales): Two gene systems control movement. Proc. Nat. Acad. Sci., U.S.A. 171:177–91.
23. Dworkin, M. and Eide, D. 1983. Mycococcus xanthus does not respond chemotactically to moderate concentration gradients. J. Bacteriol. 154:437–42.
24. Dworkin, M. 1983. Tactic behavior of Myxococcus xanthus. J. Bacteriol. 154:452–59.
25. See ref. 5(A).
26. Kaiser, D. and Dworkin, M. 1975. Gene transfer to a myxobacterium by Escherichia coli phage P1. Science 187:653–54.
27. Kuner, J. M. and Kaiser, D. 1981. Introduction of transposon Tn5 into Myxococcus for analysis of developmental and other nonselectable mutants. Proc. Nat. Acad. Sci., U.S.A. 78:425–29.
28. Orndorff, P., Stellwag, E., Starich, T., Dworkin, M., and Zissler, J. 1983. Genetic and physical characterization of lysogeny by bacteriophage Mx8 in Myxococcus xanthus. J. Bacteriol. 154:772–79.
29. Yee, T. and Inouye, M. 1981. Reexamination of the genome size of myxobacteria, including the use of a new method for genome size analysis. J. Bacteriol. 145:1257–65.
30. Morrison, C. E. and Zusman, D. R. 1979. Myxococcus xanthus mutants with temperature-sensitive, stage-specific defects: Evidence for independent pathways in development. J. Bacteriol. 140:1036–42.
31. Hagen, D. C., Bretscher, A. P., and Kaiser, D. 1978. Synergism between morphogenetic mutants of Myxococcus xanthus. Develop. Biol. 64:284–96.
32. Burchard, R. P. and Parish, J. H. 1975. Chloramphenicol resistance in Myxococcus xanthus. Antimicrob. Agents and Chemother. 7:233–38.
33. Grimm, K. and Kühlwein, H. 1973. Untersuchungen an spontanen Mutanten von Archangium violaceum (Myxobacterales): II. Uber die Einfluss des Schleims auf die Bewegung der Zellen und die Entstehung stabiler suspension Kulturen. Arch. f. Mikrobiol. 89:121–32.
34. Burchard, R. P., Burchard, A. C., and Parish, J. H. 1977. Pigmentation phenotype instability in Myxococcus xanthus. Canad. J. Microbiol. 23:1657–62.
35. Campos, J. M. and Zusman, D. R. 1978. Isolation of bacteriophage MX-4, a generalized transducing phage for Myxococcus xanthus. J. Mol. Biol. 119:167–78.
36. Martin, S., Sodergren, E., Masuda, T., and Kaiser, D. 1978. Systematic isolation of transducing phages for Myxococcus xanthus. Virol. 88:44–53.

37. See ref. 28.
38. (A) Shimkets, L. J., Gill, R. E., and Kaiser, D. 1983. Developmental cell interactions in *Myxococcus xanthus* and the spo C locus. Proc. Nat. Acad. Sci., U.S.A. *80*:1406-10.
    (B) Inouye, S., Ike, Y., and Inouye, M. 1983. Tanden repeat of the genes for protein S, a development specific protein of *Myxococcus xanthus*. J. Biol. Chem. *258*:38-40.
    (C) O'Connor, K. A. and Zusman, D. R. 1983. Coliphage P1-mediated transduction of cloned DNA from *Escherichia coli* to *Myxococcus xanthus*: Use for complementation and recombinational analysis. J. Bacteriol. *155*:317-29.
39. See ref. 28.
40. Shimkets, L. J. and Dworkin, M. 1979. Excreted adenosine is a cell density signal for the initiation of fruiting body formation in *Myxococcus xanthus*. Develop. Biol. *84*:51-60.
41. (A) Dworkin, M. 1963. Nutritional regulation of morphogenesis in *Myxococcus xanthus*. J. Bacteriol. *86*:67-72.
    (B) Manoil, C. and Kaiser, D. 1980. Guanosine pentaphosphate and guanosine tetraphosphate accumulation and induction of *Myxococcus xanthus* fruiting body development. J. Bacteriol. *141*:305-15.
42. Campos, J. M. and Zusman, D. R. 1975. Regulation of development in *Myxococcus xanthus*: Effect of 3':5'-cyclic AMP, ADP and nutrition. Proc. Nat. Acad. Sci., U.S.A. *72*:518-22.
43. Inouye, M., Inouye, S., and Zusman, D. R. 1979. Gene expression during development of *Myxococcus xanthus*: Pattern of protein synthesis. Devel. Biol. *68*:579-91.
44. Inouye, S., Inouye, M., McKeever, B., and Sarma, R. 1980. Preliminary cystallographic data for protein S, a development-specific protein of *Myxococcus xanthus*. J. Biol. Chem. *255*:3713-14.
45. Inouye, M., Inouye, S., and Zusman, D. R. 1979. Biosynthesis and self-assembly of protein S, a development specific protein of *Myxococcus*. Proc. Nat. Acad. Sci., U.S.A. *76*:209-213.
46. (A) Smith, B. A. and Dworkin, M. 1981. *Myxococcus xanthus* synthesizes a stabilized messenger RNA during fruiting body formation. Curr. Microbiol. *6*:95-100.
    (B) Nelson, D. R. and Zusman, D. R. 1983. Evidence for long-lived mRNA during fruiting body formation in *Myxococcus xanthus*. Proc. Nat. Acad. Sci., U.S.A. *80*:1467-71.
50. Dworkin, M. 1973. Cell-cell interactions in the myxobacteria. *In* Microbial differentiation, Vol. 23. Soc. Gen. Microbiol. Symp. (J. M. Ashworth and J. E. Smith, eds.). Cambridge Univ. Press, Cambridge, England, pp. 125-42.
51. See. ref. 7.
52. Lancastre, H. de and Piggot, P. J. 1979. Identification of different sites of expression for *spo* loci by transformation of *Bacillus subtilis*. J. Gen. Microbol. *114*:377-89.
53. Wireman, J. W. and Dworkin, M. 1975. Morphogenesis and developmental interactions in myxobacteria. Science *189*:516-22.

54. Oosawa, K. and Inouye, Y. 1983. Glycerol and ethylene glycol: Members of a new class of repellants of *Escherichia coli* chemotaxis. J. Bacteriol. *154*:104-12.
55. Smith, B. A. and Dworkin, M. 1980. Adenylate energy charge during fruiting body formation by *Myxococcus xanthus*. J. Bacteriol. *142*:1007-9.
56. Reichenbach, H. and Dworkin, M. 1970. Induction of myxospore formation in *Stigmatella aurantiaca* (Myxobacterales) by monovalent cations. J. Bacteriol. *101*:325-26.
57. Dworkin, M. and Voelz, H. 1967. The formation and germination of microcysts in *Myxococcus xanthus*. J. Gen. Microbiol. *28*:81-85.
58. See ref. 30.
59. Ramsey, W. S. and Dworkin, M. 1968. Microcyst germination in *Myxococcus xanthus*. J. Bacteriol. *95*:2249-57.
60. (A) Stephens, K. and White, D. Morphogenetic effects of light and guanine derivatives on the fruiting myxobacterium *Stigmatella aurantiaca*. J. Bacteriol. *144*:322-26.
    (B) White, D., Shropshire, W. Jr., and Stephens, K. 1980. Photocontrol of development by *Stigmatella aurantiaca*. J. Bacteriol. *142*:1023-24.
    (C) Qualls, G. T., Stephens, G., and White, D. 1978. Morphogenetic movements and multicellular development in the fruiting myxobacterium, *Stigmatella aurantiaca*. Devel. Biol., *66*:270-274.
    I have excluded from this generalization the effects of visible light on the development in prokaryotes of the photosynthetic apparatus.
61. For a more complete coverage of *S. aurantiaca* than is found in most reviews, see:
    (A) White, D. 1981. Cell interactions and the control of development in myxobacteria populations. Int'l Rev. Cytol. *71*:203-27.
    (B) White, D. 1984. The structure of myxobacterial cells, myxospores, and fruiting bodies. *In* Development and cell interactions in the myxobacteria (E. Rosenberg, ed.). Springer-Verlag, N.Y., pp. 51-67.
62. See refs. 23, 24.
63. Stephens, K., Hegeman, G. D., and White, D. 1982. Pheromone produced by the myxobacterium, *Stigmatella aurantiaca*. J. Bacteriol. *149*:739-47.

Chapter 8 The Relationship between Environment and Development

1. Lechevalier, H. and Holbert, P. E. 1965. Electron microscopic observation of the sporangial structure of a strain of Actinoplanes. J. Bacteriol. *89*:217-22.
2. Waterbury, J. B. 1979. Developmental patterns of pleurocapsalean cyanobacteria. *In* Developmental biology of prokaryotes (J. H. Parish, ed.). Univ. of California Press, Berkeley and Los Angeles, pp. 203-26.
3. Freese, E. 1980. Initiation of bacterial sporulation. *In* Sporulation and Germination (H. Levinson, A. L. Sonenshein, and D. J. Tipper, eds.). ASM, Washington, D.C., pp. 1-12.

4. Schaeffer, P., Millett, J., and Aubert, J. P. 1965. Catabolic repression of bacterial sporulation. Proc. Natl. Acad. Sci., U.S.A. *54*:704-11.
5. (A) Bernlohr, R. W., Haddox, M. K., and Goldberg, N. D. 1974. Cyclic guanosine 3':5'-monophosphate in *Escherichia coli* and *Bacillus licheniformis*. J. Biol. Chem. *249*:4329-31.
   (B) Setlow, P. 1973. Inability to detect cyclic AMP in vegetative or sporulating cells or dormant spores of *Bacillus megaterium*. Biochem. Biophys. Res. Commun. *52*:365-72.
6. Lopez, J. M., Ochi, K., and Freese, E. 1981. Initiation of *Bacillus subtilis* sporulation caused by the stringent response. *In* Sporulation and germination (H. Levinson, A. L. Sonenshein, and D. J. Tipper, eds.). ASM, Washington, D.C., pp. 128-133.
7. Rhaese, H. 1980. Role of the *spo* OF gene in differentiation in *Bacillus subtilis*. Ibid., pp. 134-37.
8. (A) Manoil, C. and Kaiser, D. 1980. Accumulation of guanosine tetraphosphate and guanosine pentaphosphate in *Myxococcus xanthus* during starvation and myxospore formation. J. Bacteriol. *141*:297-304.
   (B) Manoil, C. and Kaiser, D. 1980. Guanosine pentaphosphate and guanosine tetraphosphate accumulation and induction of *Myxococcus xanthus* fruiting body development. J. Bacteriol. *141*:305-15.
9. Morrison, C. E. and Zusman, D. R. 1979. Mutants of *Myxococcus xanthus* with temperature sensitive stage-specific defects: Evidence for independent pathways in development. J. Bacteriol. *140*:1036-42.
10. Fogg, G. E. 1944. Growth and heterocyst production in *Anabaena cylindrica*. Lemm. new Phytol. *43*:164-75.

Chapter 9 The Molecular Basis of Morphogenesis

1. Edelman, G. 1984. Cell-adhesion molecules: A molecular basis for animal form. Sci. Amer. *250*(4):118-129.
2. Thompson, D'Arcy W. 1961. On growth and form. An abridged edition edited by J. T. Bonner. Cambridge Univ. Press, Cambridge, England. The second edition was published in 1942.
3. Previc, E. P. 1970. Biochemical determination of bacterial morphology and the geometry of cell division. J. Theoret. Biol. *27*:471-97.
4. Daneo-Moore, L. and Shockman, G. D. 1977. The bacterial cell surface in growth and division. *In* The synthesis, assembly and turnover of cell surface components (G. Poste and G. L. Nicolson, eds.). Elsevier/N. Holland Biomedical Press, pp. 497-715. This review contains an excellent compilation and evaluation of various data that attempt to relate shape to peptidoglycan composition.
5. (A) Koch, A. L., Higgins, M. L., and Doyle, R. J. 1981. Surface tension-like forces determine bacterial shapes. J. Gen. Microbiol. *123*:151-61.
   (B) Koch, A. L., Higgins, M. L., and Doyle, R. J. 1982. The role of surface stress in the morphology of microbes. Ibid. *128*:927-45.

(C) Koch, A. L. 1982. On the growth and form of *Escherichia coli*. Ibid. *128*:2527-39.
(D) Koch, A. L. 1983. The surface stress theory of microbial morphogenesis. Adv. Microbial Physiol. *24*:301-66.
6. Iino, T. 1969. Polarity of flagellar growth in *Salmonella*. J. Gen. Microbiol. *56*:227-39.
7. (A) Asakura, S. 1970. Polymerization of flagellin and polymorphism of flagella. Adv. Biophys. *1*:99-146. A very detailed and extensive discussion of the subject.
(B) Calladine, C. R. 1976. Design requirements for the construction of bacterial flagella. J. Theor. Biol. *57*:469-89.
8. Nierhaus, K. H. 1982. Structure, assembly and function of ribosomes. Curr. Topics Microbiol. Immunol. *97*:81-155.
9. Lewis, J. C., Snell, N. S., and Burr, H. L. 1960. Water permeability of bacterial spores and the concept of a contractile cortex. Science *132*:544-45.
10. See P. Wolk's chapter "Heterocysts." *In* The biology of the cyanobacteria (N. G. Carr and B. A. Whitton, eds). Univ. of California Press, Berkeley and Los Angeles, 1982, pp. 359-68.
11. White, D., Dworkin, M., and Tipper, D. 1968. Peptidoglycan of *Myxococcus xanthus:* Structure and relation to morphogenesis. J. Bacteriol. *95*:2186-97.
12. Kottel, R. H., Bacon, K., Clutter, D., and White, D. 1975. Coats from *Myxococcus xanthus:* Characterization and synthesis during myxospore differentiation. J. Bacteriol. *124*:550-57.
13. White, D. 1984. The structure of myxobacterial cells, myxospores and fruiting bodies. *In* Development and cell interactions in the Myxobacteria (E. Rosenberg, ed.). Springer-Verlag, New York.
14. See the references to Protein S in Chapter 7.
15. Rees, D. A. 1972. Shapely polysaccharides. Biochem. J. *26*:257-73. This is a seminal article of considerable importance. For a more detailed discussion of the molecular aspects of polysaccharide shape see a slim book written by Rees: Polysaccharide Shapes. Outline Series in Biology, Wiley & Sons, 1977.

Chapter 10 Regulation of Development

1. Wright, B. E. 1973. Critical variables in differentiation. Prentice-Hall, Englewood Cliffs, N.J. See also Gunther Stent's brief essay on this subject (Ref. 5, Chap. 13).
2. Paigen, K. 1980. Temporal genes and other developmental regulators in mammals. *In* The molecular genetics of development (T. Leighton and W. F. Loomis, eds.). Academic Press, N.Y., pp. 419-70.
3. Kimchi, A. and Rosenberg, E. 1976. Linkages between deoxyribonucleic acid synthesis and cell division in *Myxococcus xanthus*. J. Bacteriol. *128*:69-79.
4. Osley, M. A. and Newton, A. 1980. Temporal control and the cell cycle in *Caulobacter crescentus:* Role of DNA chain elongation and completion. J. Molec. Biol. *138*:109-28. See also: Huguenel, E. D. and Newton, A. 1982.

Localization of surface structures during procaryotic cell differentiation: Role of cell division in *Caulobacter crescentus*. Differentiation 21:71–78.
5. Jarvik, J. and Botstein, D. 1973. A genetic method for determining the order of events in a biological pathway. Proc. Nat. Acad. Sci., U.S.A. 70:2046–50.
6. Mitchison, J. M. 1971. The biology of the cell cycle, Cambridge Univ. Press, Cambridge.
7. For an excellent review of this fundamental idea of vectorial metabolism see F. Harold's chapter "Vectorial Metabolism". *In* The bacteria: Vol. 6 (L. N. Ornston and J. R. Sokatch, eds.). Academic Press, N.Y., 1978, pp. 463–521.
8. Shapiro, L. 1983. Regulation of temporal and spatial differentiation in *Caulobacter crescentus*. *In* Microbiology 1983 (D. Schlessinger, ed.). ASM, Washington, D.C., pp. 178–82.
9. Losick, R. 1981. Sigma factors, stage 0 genes and sporulation. *In* Sporulation and germination (H. Levinson, A. L. Sonenshein, and D. J. Tipper, eds.). ASM, Washington, D.C., pp. 48–56.
10. Trempy, J. E., Morrison-Plummer, J., and Haldenwang, W. G. 1984. Synthesis of σ^{29}, an RNA polymerase specificity determinant, is a developmentally regulated event in *Bacillus subtilis*. J. Bacteriol. 161:340–46.
11. See Chapter 5, ref. 15.
12. Amemiya, K. and Shapiro, L. 1977. *Caulobacter crescentus* RNA polymerase: Purification and characterization of holoenzyme and core polymerase. J. Biol. Chem. 252:4157–65.
13. Rudd, K. and Zusman, D. R. 1982. RNA polymerase of *Myxococcus xanthus*: Purification and selective transcription with bacteriophage templates. J. Bacteriol. 151:89–105.
14. Westpheling, J., Ranes, M., and Losick, R. 1984. Polymerase heterogeneity in *Streptomyces*. Nature 313:22–27.
15. Jenkinson, H. F., Kay, D., and Mandelstam, J. 1980. Temporal dissociation of late events in *Bacillus subtilis* sporulation from expression of genes that determine them. J. Bacteriol. 141:793–805.
16. Tipper, D. J., Johnson, C. W., Ginther, C. L., Leighton, T., and Wittman, H. G. 1977. Erythromycin-resistant mutations in *Bacillus subtilis* cause temperature sensitive sporulation. Mol. Gen. Genet. 150:147–59.
17. (A) Smith, B. A. and Dworkin, M. 1981. *Myxococcus xanthus* synthesizes a stabilized messenger RNA during fruiting body formation. Cur. Microbiol. 6:95–100.
    (B) Nelson, D. R. and Zusman, D. R. 1983. Evidence for long-lived mRNA during fruiting body formation in *Myxococcus xanthus*. Proc. Nat. Acad. Sci., U.S.A. 80:1467–71.
18. Piggot, P. J., Moir, A., and Smith, D. A. 1981. Advances in the genetics of *Bacillus subtilis* differentiation. *In* Sporulation and germination (H. Levinson, A. L. Sonenshein, and D. J. Tipper, eds.). ASM, Washington, D.C., pp. 29–39.
19. Nevers, P. and Saedler, H. 1977. Transposable genetic elements as agents of gene instability and chromosomal rearrangements. Nature 268:109–15.
20. Zeig, J., Silverman, M., Hilman, M., and Simon, M. 1977. Recombinational switch for gene expression. Science 196:170–72. Also see an excellent review on the genetics of phase variation: Silverman, M. and Simon, M. 1980. Phase variation: Genetic analysis of switching mutants. Cell 19:845–54.
21. See Chapter 9, reference 1.

Chapter 11 Exchange of Signals

1. Horgen, P. A. 1979. Steroid induction of differentiation in *Achlya* as a model system. *In* Eukaryotic microbes as model developmental systems (D. H. O'Day and P. A. Horgen, eds.). Dekker, N.Y., pp. 272-93.
2. Newell, P. C. 1977. Aggregation and cell surface receptors in cellular slime molds. *In* Microbial interactions (J. L. Reissig, ed.). Chapman and Hall, London, pp. 1-57. This is an excellent and comprehensive overview of development in *Dictyostelium discoideum* by one of the leading investigators in the field.
   For a background on the biology of the organism see: Loomis, W. F. (ed.). 1982. The development of *Dictyostelium discoideum*. Academic Press, N.Y. This book was preceded by the first version, *Dictyostelium discoideum: A developmental system*. Academic Press, 1975.
3. (A) Keller, H. V. 1981. The relationship between leucocyte adhesion to solid substrata, locomotion, chemokinesis and chemotaxis. *In* Biology of the chemotactic response (J. M. Lackie and P. C. Wilkinson, eds.). Cambridge Univ. Press, Cambridge, pp. 27-51. This is an excellent general description of leucocyte migration.
   (B) Wilkinson, P. C. 1981. Peptide and protein chemotactic factors and their recognition by neutrophil leucocytes. Ibid., pp. 53-72. This work emphasizes the effects of the defined peptide and protein chemo-attractants for leucocytes.
   (C) Zigmond, S. H. and Sullivan, S. J. 1981. Receptor modulation and its consequences for the response to chemotactic peptides. Ibid., pp. 73-87. This article is concerned mainly with the interactions between chemotactic peptides and their leukocyte receptors.
4. Black, I. B. 1979. Neuronal responses to extracellular signals. *In* The role of intercellular signals, navigation, encounter, outcome (J. G. Nicholls, ed.). Dahlem Konferenzen, Berlin, pp. 155-78. This work is an excellent review of the remarkable variety of neuronal responses to extracellular signals.
5. Cuatrecasas, P. and Hollenberg, M. D. 1976. Membrane receptors and hormone action. Adv. Prot. Chem. 30:251-51. This work presents a review of hormone-receptor interactions and the effect of these interactions on membrane functions and cell behavior.
6. Manney, T. R., Duntze, W., and Retz, R. 1981. The isolation, characterization, and physiological effects of the *Saccharomyces cerevisiae* sex pheromones. *In* Sexual interactions in eukaryotic microbes (D. H. O'Day and P. A. Horgen, eds.). Academic Press, N.Y., pp. 21-51.
7. Horgen, P. 1981. The role of the steroid sex pheromone antheridiol in controlling the development of male sex organs in the water mold, *Achyla*. Ibid., pp. 155-78.
8. Burger, M. M. 1977. Mechanisms of cell-cell recognition: Some comparisons between lower organisms and vertebrates. *In* Cell interactions in differentiation (M. Karkinen-Jääskeläinen, L. Saxén, and L. Weiss, eds.). Academic Press, London, pp. 357-76.
9. Goodenough W. W. 1977. Mating interactions in *Chlamydomonas*. *In* Microbial interactions (J. L. Reissig, ed.). Chapman and Hall, London, pp. 325-50.

10. (A) Gerisch, G., Krelle, H., Bozzars, S., Eitle, E., and Guggenheim, R. 1980. Analysis of cell adhesion in *Dictyosteliun* and *Polysphondylium* by the use of Fab. In Cell adhesion and motility (A. Curtis and J. Pitts, eds.). Cambridge Univ. Press, Cambridge, pp. 293–307. This review is mainly concerned with the nature and role of the surface glycoprotein referred to as contact sites A.
    (B) Barondes, S. H. 1980. Developmentally regulated lectins in slime molds and chick tissues: Are they cell adhesion molecules? Ibid., pp. 309–28. This work covers the nature and role of a set of lectins, called *discoidins* isolated from *D. discoideum* and postulated to be involved in developmental adhesion.
11. Cantor, H. and Gershon, R. K. 1979. Cellular communication among immunological cells. In The role of intercellular signals, navigation, encounter, outcome (J. G. Nicholls, ed.). Dahlem Konferenzen, Berlin, pp. 55–74.
12. Spenser, E. The fairie queene; book II, canto 12, stanzas 72, 77.
13. The original paper, describing discoidin, was by Rosen, S. N., Kaflea, J. A., Simpson, D. L., and Barondes, S. H. 1973. Developmentally regulated carbohydrate-binding protein in *Dictyostelium discoideum*. Proc. Natl. Acad. Sci. U.S.A. 70:2554–57. Recently, Barondes's laboratory has shown that discoidin I shares a four-amino acid sequence with the adhesion protein fibronectin and proposes that discoidin I functions as a cell-substrate adhesin and plays a role in the ability of the organism to form streams of cells (Springer, W., R., Cooper, D. N. W., and Barondes, S. H. 1984. Discoidin I is implicated in cell-substratum attachment and ordered cell migration of *Dictyostelium discordeum* and resembles fibronectin. Cell 39:557–564).

    The work that describes the discoidin-minus mutants that nevertheless develop normally is: Alexander, S., Schinnick, T. M., and Lerner, R. A. 1983. Mutants of *Dictyostelium discoideum* blocked in expression of all members of the developmentally regulated discoidin multigene family. Cell 34:467–75. Finally, the paper describing the absence of discoidin from the cell surface is: Erdos, G. W. and Whitaker, D. 1983. Failure to detect immunocytochemically reactive endogenous lectin on the cell surface of *Dictyostelium discoideum*. J. Cell. Biol. 97:990–1000.
14. Machlis, L. 1972. The coming of age of sex hormones in plants. Mycologia 65:235–47.
15. See J. Thorner (1980. Intercellular interactions of the yeast *Saccharomyces cerevisiae*. In The molecular genetics of development [T. Leighton and W. F. Loomis, eds.]. Academic Pressi, N.Y., pp. 119–78) for an excellent discussion of *α* and "a" factors.
16. Herskowitz, I., Blair, L., Forbes, D., Hicks, J., Kasser, Y., Kushner, P., Rine, J., Sprague, G. Jr., and Strathern, J. 1980. Control of cell type in the yeast *Saccharomyces cerevisiae* and a hypothesis for development in higher eukaryotes. Ibid., pp. 79–118.
17. Horgen, P. A. 1977. Steroid induction of differentiation: *Achyla* as a model system. In Eukaryotic microbes as model developmental systems (D. H. O'Day and P. A. Horgen, eds.). Dekker, pp. 272–94.
18. Dunny, C. M., Brown, B. L. and Clewell, D. B. 1978. Induced cell aggregation and mating in *Streptococcus faecalis*: Evidence for a bacterial sex hormone. Proc. Nat. Acad. Sci., U.S.A. 75:3479–83.

19. Neilson, K. H., and Hastings, J. W. 1979. Bacterial bioluminescence: Its control and ecological significance. Microbiol. Rev. *43*:406-518.
20. See the recent review by M. Dworkin and D. Kaiser (1985. Cell interactions in myxobacterial growth and development. Science, in press).
21. Kaiser, D. 1979. Social gliding is correlated with the presence of pili in *Myxococcus xanthus*. Proc. Nat. Acad. Sci., U.S.A. *76*:5952-56.
22. Achtman, M. and Skurray, R. 1977. A redefinition of the mating phenomenon in bacteria. *In* Microbial interactions (J. L. Reissig, ed.). Chapman and Hall, London, pp. 233-80.
23. Rosenberg, E., Keller, K. H., and Dworkin, M. 1977. Cell density-dependent growth of *Myxococcus xanthus* on casein. J. Bacteriol. *129*:770-77.
24. Ramsey, W. S. and Dworkin, M. 1968. Microcyst germination in *Myxococcus xanthus*. J. Bacteriol. *95*:2249-57.
25. (A) Wireman, J. W. and Dworkin, M. 1975. Morphogenesis and developmental interactions in Myxobacteria. Science *189*:516-23.
    (B) Shimkets, L. J. and Dworkin, M. 1981. Excreted adenosine is a cell density signal for the initiation of fruiting body formation in *Myxococcus xanthus*. Devel. Biol. *84*:51-60.
26. Nelson, D. R., Cumsky, M. G., and Zusman, D. R. 1981. Localization of myxobacterial hemagglutinin in the periplasmic space and on the cell surface of *Myxococcus xanthus* during developmental aggregation. J. Biol. Chem. *265*:12589-95.
27. Stephens, K., Hegeman, C.D., and White, D. 1982. Pheromone produced by the myxobacterium, *Stigmatella aurantiaca*. J. Bacteriol. *149*:739-747.

Chapter 12 Genetic Approaches to Studying Development

1. Botstein, D. and Maurer, R. 1982. Genetic approaches to the analysis of microbial development. Ann. Rev. Genetics *16*:61-83. See also Piggot's review (Chapter 3, ref. 10[B]) for a very useful description of genetic approaches to the analysis of development.
2. Stanier, R. Y. and Cohen-Bazire, G. 1957. The role of light in the microbial world: Some facts and speculations. *In* Microbial ecology, 7th Symp. Soc. Gen. Microbiol. (R.E.D. Williams and C. C. Spicer, eds.). Cambridge Univ. Press, London.
3. (A) Drake, J. W. 1969. Mutagenic mechanisms. Ann. Rev. Genet. *3*:247-68.
   (B) Drake, J. W: 1970. The molecular basis of mutation. Holden-Day, San Francisco.
4. (A) Kleckner, N., Roth, J., and Botstein, D. 1977. Genetic engineering *in vivo* using translocatable drug resistance elements: New methods in bacterial genetics. J. Molec. Biol. *116*:125-259.
   (B) Kuner, J. M., Avery, L., Berg, E. D., and Kaiser, D. 1981. Use of transposon Tn5 in the genetic analysis of *Myxococcus xanthus*. *In* Microbiology 1981 (D. Schlessinger, ed.). ASM, Washington, D.C., pp. 128-32.
5. Jarvick, J. and Botstein, D. 1973. A genetic method for determining the order of events in a biochemical pathway. Proc. Nat. Acad. Sci., U.S.A. *70*:2046-50.

6. Morris, N. R., Lai, M. H., and Oakley, C. E. 1979. Identification of a gene for α-tubulin in *Aspergillus nidulans*. Cell *16*:437–42.
7. (A) Kuner, J. M. and Kaiser, D. 1981. Introduction of transposon Tn5 into *Myxococcus* for analysis of developmental and other non-selectable mutants. Proc. Nat. Acad. Sci., U.S.A. *78*:425–29.

    (B) Purecker, M., Bryan, R., Amemiya, K., Ely, B., and Shapiro, L. 1982. Isolation of a *Caulobacter* gene cluster specifying flagellin production by using nonmotile Tn5 insertion mutants. Proc. Nat. Acad. Sci., U.S.A. *79*:6997–6801.

    It is now possible to insert the *Streptococcus* transposon Tn917 into *B. subtilis* via the phage SPβ. Zahler, S. A., Korman, R. A., Odebralski, J. M., Fink, P. S., Mackey, C. J., Poutre, C. G., Lipsky, R., and Youngman, P. J. 1982. Genetic manipulation with phage SPβ. *In* Molecular cloning and gene regulation in bacilli (A. T. Ganesan, J. Chang, and J. A. Hoch, eds.). Academic Press, New York.
8. See ref. 7(A).
9. Casadoban, M. J. and Cohen, S. N. 1979. Lactose genes fused to exogenous promoters in one step using a Mu-lac bacteriophage: *In vivo* probe for transcriptional control sequences. Proc. Nat. Acad. Sci., U.S.A. *76*:4530–33.
10. Inouye, S., Ike, Y., and Inouye, M. 1983. Tandem repeat of the genes for protein S, a development specific protein of *Myxococcus xanthus*. J. Biol. Chem. *258*:38–40.
11. Shimkets, L. J., Gill, R. E., and Kaiser, D. 1983. Developmental cell interactions in *Myxococcus xanthus* and the *Spo C* locus. Proc. Nat. Acad. Sci., U.S.A. *80*:1406–10.

Chapter 13 Epilogue

1. O'Day, D. H. and Horgen, P. A. (eds.). 1979. Eucaryotic microbes as model developmental systems. M. Dekker, Inc., N.Y. This book is a collection of individually authored chapters on various aspects of development in protozoa, fungi, yeast, slime molds, and algae.
2. See Chapter 11, ref. 2, Loomis entry.
3. Yamada, K. M. 1983. Cell interactions and development. J Wiley & Sons, N.Y. This work is another multiauthored book including chapters on various aspects of interactions among developing microbes (yeast, *Chlamydomonas, Dictyostelium, Rhizobium*-plant) and among higher organisms.
4. See ref. 5(F), Chap. 7.
5. Stent, G. 1985. Thinking in one dimension: The impact of molecular biology on development. Cell *40*:1–2.

Figure References and Credits

Figure 2-1 Tortora, G. J., Funke, B. R., and Case, C. L. 1982. Microbiology. An introduction. Benjamin/Cummings Publishing Co., p. 325, Fig. 12–13.

References

Figure 2-2   Mandelstam, J., McQuillen, K., and Dawes, I. (eds.) 1982. Biochemistry of bacterial growth, 3rd edition. Blackwell Scientific Publications, p. 373, Fig. 14.
Figure 2-3   Lewin, B. 1983. Genes. Wiley Publishing Co., p. 268, Fig. 16.8.
Figure 2-4   Cell differentiation: Molecular basis and problems. Nover, Luckner, L. M., Parthier, B., (eds.) 1982. Springer-Verlag, p. 123, Fig. 4.7.
Figure 2-5   Courtesy of Dr. M. Simon.
Figure 2-6   Mandelstam, J., McQuillen, J., and Dawes, I. (eds.) 1982. Biochemistry of bacterial growth, 3rd edition. Blackwell Scientific Publications, p. 370, Fig. 12.
Figure 3-1   Courtesy of Dr. S. C. Holt.
Figure 3-2   Mandelstam, J. and McQuillen, K. (eds.) 1968. Biochemistry of Bacterial growth. Blackwell Publishing Co., p. 468, Fig. 2.
Figure 3-4   Courtesy of Dr. S. C. Holt.
Figure 3-5   Tipper, D. et al. 1977. *In* Microbiology 1977 (D. Schlessinger, ed.). American Society for Microbiology, pp. 50-68.
Figure 3-7   Wilkenson, B. J. et al. 1975. Biochem. J. *152*:561-69, Fig. 1.
Figure 3-8   Dawes, I. W. and Hansen, J. N. 1972. Crit. Rev. Microbiol. *1*:479-520, Fig. 8.
Figure 3-9   Strange, R. E. and Hunter, J. R. 1969. *In* The bacterial spore (G. W. Gould and A. Hurst, eds.). Academic Press, p. 461, Fig. 4.
Figure 3-10   Henner, D. J. and Hoch, J. A. 1980. Microbiol. Rev. *44*:57-82, Fig. 6.
Figure 4-1   Shapiro, L. et al. 1981. *In* Soc. Gen. Microbiol. Sympos. No. 31, p. 317-39, Fig. 1.
Figure 4-2   Poindexter, J. S. 1964. Bact. Rev. *28*:231-95, Fig. 7a.
Figure 4-3   Schmidt, J. M. and Stanier, R. Y. 1966. J. Cell. Biol. *28*:423-36, Figs. 4, 5. By copyright permission of the Rockefeller University Press.
Figure 4-4 (A)   Johnson, R. C. et al. 1979. J. Bacteriol. *138*:984-89, Fig. 2.
(B)   Ibid., Fig. 3.
Figure 4-5   Poindexter, J. S. 1981. Microbiol. Rev. *45*:123-79, Fig. 4-C.
Figure 4-6   Ibid., Fig. 5.
Figure 4-7   Shapiro, L. 1976. Ann. Rev. Microbiol. *30*:377-407, Fig. 2.
Figure 4-8   Huguenel, E. D. and Newton, A. 1983. Diff. *21*:81-79, Fig. 5c.
Figure 4-9   Shapiro, L. et al. 1981. *In* Soc. Gen. Microbiol. Sympos. No. 31, p. 317-39, Fig. 3.
Figure 5-1   Stanier, R. Y. and Cohen-Bazire, G. 1977. Ann. Rev. Microbiol. *31*:225-74, Fig. 8.
Figure 5-2   Wilcox, M. et al. 1975. *In* Microbiology 1975 (D. Schlessinger, ed.). American Society for Microbiology, p. 453-63, Fig. 11.
Figure 5-3   Lang, N. 1968. *In* Algae, man and the environment (D. F. Jackson, ed.). Syracuse University Press, pp. 235-48.
Figure 5-4   Stanier, R. Y. and Cohen-Bazire, G. 1977. Ann. Rev. Microbiol. *31*:225-74, Figure 9.
Figure 5-5   Fay, P. and Lang, N. J. 1971. Proc. Royal Soc. Lond. Sec. B *178*:185-92.
Figure 5-6   Wolk, P. 1979. *In* Determinants of spatial organization (S. Subtelny and I. R. Konigsberg, eds.). Academic Press, p. 247-66.
Figure 5-7   Wilcox, M. 1975. *In* Microbiology 1975 (D. Schlessinger, ed.). American Society for Microbiology, p. 453-63, Fig. 5.

References

Figure 6-1 Lechevalier, H. and Holbert, P. E. 1965. J. Bacteriol. 89:217–22.
Figure 6-2 Ibid., Figs. 9–15.
Figure 6-3 Courtesy of Dr. H. Lechevalier.
Figure 6-4 Chater, K. E. and Hopwood, D. A. 1973. In Microbial differentiation, Soc. Gen. Microbiol. Symp. No. 23 (J. M. Ashworth and J. E. Smith, eds.). Academic Press, pp. 143–60, Fig. 1.
Figure 6-5 Ibid., Plate 1.
Figure 6-6 Ibid., Plate 2.
Figure 6-7 McVittie, A. 1974. J. Gen. Microbiol. 81:291–302, Fig. 3b.
Figure 6-8 Wildermuth H. and Hopwood, D. A., 1970. J. Gen. Microbiol. 60:51–59, Fig. 2.
Figure 6-9 McVittie, A. 1974. J. Gen. Microbiol. 81:291–302, Fig. 1a.
Figure 6-10 Ibid., Fig. 1d.
Figure 6-11 Ibid, Figs. 4a and 4b.
Figure 6-12 Hopwood, D. A. 1976. In Handbook of biochemistry and molecular biology: Nucleic acids, Vol. 2 (G. D. Fasman, ed.). CRC Press, pp. 723–28.
Figure 6-13 Hopwood, D. A. et al. 1973. Bacteriol. Rev. 37:371–405, Fig. 11.
Figure 6-14 Merrick, M. J. 1976. J. Gen. Microbiol. 96:299–315, Fig. 2.
Figure 7-1 Thaxter, R. 1892. Botan. Gaz. 17:389–406, Plates 22, 23.
Figure 7-2 Thaxter, R. 1897. Botan. Gaz. 23:395–411, Plates 30, 31.
Figure 7-5 Reichenbach, H. and Dworkin, M. 1981. In The Prokaryotes, Vol. I (M. P. Starr et al., eds.).. Springer-Verlag, Fig. 7f.
Figure 7-6 Courtesy of Dr. Hans Reichenbach.
Figure 7-7 Ibid.
Figure 7-8 Ibid.
Figure 7-9 Ibid.
Figure 7-10 Reichenbach, H. and Dworkin, M. 1981. In The Prokaryotes, Vol. I (M. P. Starr et al., eds.). Springer-Verlag, Fig. 7e.
Figure 7-11 Courtesy of Dr. Hans Reichenbach.
Figure 7-12 Courtesy of Dr. Herbert Voelz.
Figure 7-13 Kaiser, D. 1979. Proc. Nat. Acad. Sci., U.S.A. 75:5952–56, Fig. 1.
Figure 7-14 Inouye, M., Inouye, S., and Zusman, D. R. 1979. Biosynthesis and self-assembly of protein S, a development-specific protein of Myxococcus xanthus. Proc. Nat. Acad. Sci. U.S.A., 76:209–13, Figs. 6a, 6c.
Figure 7-15 Reichenbach, H. Dworkin, M. 1981. In The Prokaryotes, Vol. I (M. P. Starr et al., eds.). Springer-Verlag, Fig. 2b.
Figure 7-16 White, D. 1981. Intl Rev. Cytol. 72:203–77, Fig. 2.
Figure 7-17 Kuner, J. and Kaiser, D. 1981. Prod. Nat. Acad. Sci., U.S.A., 78:425–29, Fig. 6.
Figure 7-18 Orndorff, P. 1981. The cell surface of Myxococcus xanthus. Doctoral thesis, University of Minnesota, Fig. 2.
Figure 7-19 White, D. 1975. In Spores VI (P. Gerhardt et al., eds.). American Society for Microbiology, pp. 44–51, Fig. 2.
Figure 7-20 Voelz, H. 1966. Archiv. f. Mikrobiol. 55:110–15, Figs. 1, 4.
Figure 8-1 Zusman, D. 1984. In Myxobacteria: Development and cell interactions (E. Rosenberg, ed.). Springer-Verlag, p. 185–213.
Figure 8-2 Carr, N. G. 1979. In Developmental biology of prokaryotes (J. H. Parish, ed.). University of California Press, pp. 167–201, Fig. 9-1.

Figure 8-3   Shimkets, L. and Dworkin, M. 1981. Develop. Biol. *84*:51-60, Fig. 1.
Figure 9-1   Ghuysen, J. M. 1968. Bacteriol. Rev. *32*:425-64.
Figure 9-2   Koch, A. et al. 1981. J. Gen. Microbiol. *123*:151-61, Fig. 9.
Figure 9-3   Rees, D. A. 1977. Polysaccharide shapes. Halstead Press, Fig. 5.5.
Figure 10-1 Huguenel, E. D. and Newton, A. 1982. Diff. *21*:71-78, Fig. 7a.
Figure 10-2 Losick, R. 1981. *In* Sporulation and germination (Spores VIII) (H. S. Levinson et al., eds.). American Society for Microbiology, pp. 48-56, Fig. 3.
Figure 10-3 Piggot, P. J. et al. 1981. *In* Sporulation and germination (Spores VIII) (H. S. Levinson et al., eds.). American Society for Microbiology, pp. 29-39, Fig. 2.
Figure 11-1 Courtesy of Dr. P. Newell.
Figure 11-2 Ibid.
Figure 11-3 Thorner, J. 1980. *In* The Molecular genetics of development (T. Leighton and W. F. Loomis, eds.). Academic Press, pp. 119-78, Fig. 1.
Figure 11-4 Alexopoulos, C. J. 1962. Introductory mycology, 2nd ed. John Wiley, Fig. 53.
Figure 12-1 Hodgkin, J. and Kaiser, D. 1979. Molec. Gen. Genet. *171*:167-76, Fig. 1.

Index

Achlya:
    antheridiol, 196
    mating, 196–197
    oogoniol, 197
    pheromones, 196–197
Acrasin:
    Dictyostelium discoideum, 188
    Spencer's "Fairie Queene," 188
Actinoplanes:
    development, 86–87, Figs. 6-1, 6-2, 6-3
    environment, role in development, 153, 156
Adhesins, *Dictyostelium discoideum*, 191
Adhesion, cell:
    Dictyostelium discoideum, 189–191
    Stigmatella aurantiaca, 147
"a" factor of yeast, 193
Aggregation:
    cAMP, in *Dictyostelium discoideum*, 188
    Dictyostelium discoideum, role of acrasin in, 188
    Myxobacteria, 110, 132, 136–138, 140–141, 144, 147, Figs. 7-3, 7-4, 7-18
Agmenellum, 83
Alpha factor of yeast, structure, 193
Anabaena, 69, 72–80, 83; dinitrogen-fixing genes, 83
Anacystis, 83
Antheridiol:
    mating in *Achlya*, 196
    mechanism of, 197
    structure, 196

Bacillus:
    bacteriophage, 41–43; lysogeny by SPβ, 42–43
    endospore formation, 30–31
    environment, role in development, 43–45, 155
    genetic map, 36, Fig. 3-10
    genetics, 34–41
        complementation by protoplast fusion, 40
        mutant studies, 36–39
        plasmids and recombinant DNA, 41
        protoplast fusion, 40
        transformation, 39
        transposon Tn 917, insertion of, 43
        viral transduction, 40
    genetic techniques available, 35
    growth and nutrition, 23
    life cycle, 30–34
    mutants:
        conditional, 38–39
        developmental, 37–38
        growth/DNA replication, 38
        physiological/biochemical, 38
    spore germination, 31–34
    transformation, DNA uptake, mechanism, 39
Bacteriophage, 8–21; nature of development, 8
Bacteriophage development:
    lysogeny and productive infection, choice between, 13
    transcriptional control:
        antiterminator, 11
        negative RNA polymerase modifications, 12

Bacteriophage development *(continued)*
    new RNA polymerase, 9
    positive RNA polymerase modifications, 12
Bioluminescence, autoinducer structure, 100
Bonner, J. T., 216
Botstein, D., 204

cAMP, role in development:
    Dictyostelium discoideum, 188
    endospore formation, 44, 154
    fruiting body formation in *Myxococcus xanthus*, 138
Caulobacter:
    bacteriophage, 61–62
    cell structure, 53–59
    cell surface, nature of, 53–54
    development:
        spatial orientation, 66
        temporal regulation, 67
    ecology and distribution, 52–53
    environment, role in development, 52–53, 153, 156, 158–159
    flagellin, 55
    flagellum:
        basal body, 55, Figs. 4-4, 4-5
        fate during development, 54–55
        proteins, 55
        synthesis, 65
    genetic map, 64
    genetics, 62–65
        chromosome mobilization, 63
        conjugal mating, 62
        mutants, 63
        plasmid transfer, 63
        transcriptional control, 63–64
        translational control, 63–65
        transposon mutagenesis, 63
    growth, relation to development, 66
    life cycle, 50, 51, Fig. 4-1
        DNA synthesis during, 59
        flagellin synthesis during, 60
        pilin synthesis during, 61
    methyl-accepting chemotaxis proteins (MCP), 65
    outer membrane composition, 53–54
    phage, role of genetics in, 62
    physiological correlates of development, 59–61, Fig. 4-7
    pili, 56; phage attachment, 56
    pilin, 56
    RNA polymerase, 63–64

Caulobacter (continued)
    stalk, 56–59
        composition, 56–57
        crossbands, 58–59
        function, 57
    taxonomy and natural relations, 53
Cell-cell interactions, 186–203
    bioluminescence, 199–200
    Dictyostelium discoideum:
        antibodies against cell surface antigens, 191
        contact sites A, 191
    Myxobacteria, 146–147, 200–203
    prokaryotes, 197–203
    streptococci, 199
    water molds, 196–197
    yeast, 192–196
Cell communication and interactions, definition, 6
Cell density, effect on myxobacteria, 96, 137, 141, 145, 200–202
Cellular differentiation, definition, 5
Cellular morphogenesis, definition, 5
Chemotaxis:
    Dictyostelium discoideum, 188
    myxobacteria, 129–130
CIA (clumping inducing factor), pheromone in *Streptococcus*, 199
Cloning, developmental genes, examples, 214
Cohen, S. N., 97
Cohn, Ferdinand, 22
Communication, cell, definition, 6
Complementation analysis, 208
Contact sites A:
    Dictyostelium discoideum in, 191
    properties, 191
Cyanobacteria, 68–84
    Calvin cycle, 71, 79
    carbon metabolism, 71
    carboxysome, 71
    genetics, shuttle vector, 83
    glutamate generation, 71
    heterocyst, 72–84
    nitrogen metabolism, 71
    photosynthetic metabolism, 71
    photosystems I and II, 71
    phycobiliprotein, 69
    phycobilisome, 69
    ribulose biphosphate, 71, 79
    ribulose biphosphate carboxylase, 71, 79
    thylakoids, 69, Fig. 5-1
    vegetative cells, 69–71; surface layers, 69
Cyanobacterial heterocysts, role of environment in development, 76, 77–79,

Cyanobacterial heterocysts *(continued)*
81, 157–158, Fig. 8-2
Cyclic development, 5
Cytoplasmic membrane, changes during development:
Caulobacter, 178
endospore formation, 30, Figs. 3-6, 3-7

Dancer, B., 40
Daneo-Moore, L., 165
Development, bacteriophage, 8–21
Development, prokaryotic:
alternative to growth, 2
categories, 3–6
cyclic vs. noncyclic, 4, 5
definitions, 2, 3
pleomorphism and monomorphism, 1
unicellular vs. multicellular, 4, 5
Dictyostelium discoideum:
acrasin, 188
cAMP as signal, 188
cell interactions, 188–192
contact sites A, 191
Spencer's "Fairie Queene," 188
Differentiation, cellular, definition, 5
Differentiation, population, definition, 6
Dipicolinic acid, 25, 38, 49, Fig. 3-3
Discoidins:
Dictyostelium discoideum, 192
properties, 192
role, 192
Dispersal, alternating with aggregation, 156
Dispersal mechanisms:
swarmer cells of *Caulobacter,* 51–53, 156, 158–159
zoospores of *Actinoplanes,* 86–87, 156, Figs. 6-1, 6-2, 6-3
DNA synthesis, role in development:
Caulobacter, 59, 65, 67, Fig. 4-7
endospore germination, 34, Fig. 3-8
myxospore formation, 170
myxospore germination, 145
Dow, C. S., 4

Efficient feeding cells, formation:
Caulobacter, 52–53, 158–159
stalk formation by *Caulobacter,* 51–53, 159
Endospore:
antibiotics and sporulation, 39
calcium dipicolinate, 25
chromosome, 24–25
dipicolinic acid, 25, Fig. 3-3

Endospore *(continued)*
genetics, 34–41
map, Fig. 3-10
mutant studies, 36–39
plasmids, 41
protoplast fusion, 110
recombinant DNA, 41
spo mutants, 37
techniques available, 35
transformation, 39
viral transduction, 40
germination:
activation, 31–32
initiation, 32–33
outgrowth, 33
physiological events during, 33–34, Table 3-1, Fig. 3-8
proteins, 25
stages, 31
heat resistance:
cortical expansion, contraction and water content, 49
DPA, role, 49
history, 22
resistance, mechanism, 48–49
structure, 23–29
cortex, 28
exosporium, 29–30
germ cell wall, 27–28
inner forespore membrane (IFsM), 26
outer forespore membrane (OFsM), 28, 30
protoplast, 23–25
spore coat, 28–29
Endospore formation:
cAMP, 44
commitment, 45
endotrophic sporulation, 45
induction, 43–45; multiple-step hypothesis, 44
metabolic reactions during, 46
morphological events, sequence, 30–31, Fig. 3-6
mRNA synthesis, 45
nutritional effects, 43–44
outer forespore membrane (OFsM), reversal of polarity, 30, Fig. 3-7
regulation, 45–48
antibiotics and ribosomal modifications, 48
initiation factors (IF), role, 48
sigma factors for RNA polymerase, 46–47
transcriptional control, 46–47
translational control, 47–48
triggers, 43

Environment, role in development, 152–161
Epistasis:
  Bacillus sporulation, 210
  fruiting body formation in *Myxococcus xanthus*, 210
  Streptomyces development, 97
Exosporium, 29–30
Expression of developmental genes, *lacZ* fusions, 213
Extracellular complementation, myxobacteria, 200–201

Flagella:
  phase variation:
    hin (H inversion), 16
    Salmonella, 15–18; mechanism, 17, Fig. 2-4
  self-assembly, 167–168
Foster, J. W., 45
Freese, E., 44
Fritsch, F. E., 84
Fruiting bodies, formation, myxobacteria, 159–161, Fig. 8-3

Gerisch, G., 189
Glutamine synthetase (GS)
  Anabaena, 81
  regulation in heterocyst, 81
Glycoproteins, cell surface, *Dictyostelium discoideum*, 191

Haselkorn, R., 83, 180
Henrici, A. T., 50
Herskowitz, I., 194
Heterocyst:
  anoxygenic photosynthesis, 77
  biosynthetic limitations, 77
  CO_2 fixation, absence in, 79
  CO_2 incorporation into, 79
  chlorophyll, reduced, 77
  chlorophyll P700, 77
  composition, 73–77
  dead-end cell, 77
  dinitrogen fixation, 77–79; strategy, 83
  DNA, 76
  function, 77–79
  genetics, 83
  glucose-6-phosphate dehydrogenase, increase, 79
  glycolipids, physical barrier to oxygen, 78
  GS/GOGAT, 78

Heterocyst *(continued)*
  Hill reaction, absence, 77
  isolation, 72–73
  metabolism, 77–79
  morphogenesis, 79–81
  NADPH/NADP ratio, 79
  oxygen, insulation from, 77
  photosystem II, absence, 77
  phycobiliprotein, absence, 77
  phycobilisome, absence, 77
  pigments, 76
  polysaccharides, on surface, 76
  population differentiation, 83
  6-phosphogluconate dehydrogenase, increase in, 79
  spacing, 81–83
    cell division, 81
    diffusible inhibitor, 81–82
    rules for, 81, Fig. 5-7
  structure, 73–77
  surface layers, 76; oxygen protection, 76, 170
  vegetative cell interaction with, diffusion of ^{14}C, 79
  vegetative cell relation to, 73–75, Figs. 5-3, 5-4
    microplasmodesmata, 74
    cyanophycin granules, 75, Fig. 5-4
Heteroduplex mapping, DNA flip-flop in phage mu and *Salmonella*, 17–18, Fig. 2-5
Hin (H inversion), flagella phase variation, role, 16
Hopwood, D., 40, 85, 97
Horgen, P., 197

Inouye, M., 122, 132
Interactions, cell, 186–203; definition, 6
Interaction suppressors:
  cell interactions, 211
  tubulin genes, 211
  self-assembly, 211

Jacob, François, 1
Jahn, E., 105
Jarvick-Botstein method for order of gene function, 209–210, Table 12-1

Kaiser, D., 137, 146, 207, 213
Koch, A., 167–168
Koch, Robert, 22

lacZ fusions, expression of developmental genes, measure of, 213
lacZ genes, S protein, expression in *Myxococcus xanthus*, 213
Lectins, *Dictyostelium discoideum* discoidins, 192
Lederberg, J., 2
Losick, R., 179
Luminescent bacteria, autoinducer of luminescence, 199–200
Lysogeny, choice of, 13

MacClintock, B., 184
Mandelstam, J., 182
Maurer, R., 204
Metabolically specialized cells, formation:
　dinitrogen fixation by heterocysts, 77–79, 157
　heterocyst formation, 68–84, 157
　heterocyst induction by removal of fixed nitrogen, 76, 81, 157–158, Fig. 8-2
Mitchell, P., 177
Morphogenesis, bacteriophage, 18–21
Morphogenesis, cellular, definition, 5
Morphogenesis, molecular basis, 162–174
　cyanobacterial heterocysts, 73–77, 79–81, 169–170
　Bacillus endospore, 23–29, 168–169
　flagellar self-assembly, 167–168
　fruiting body, 172, Fig. 9-3
　myxospores:
　　protein S, 172
　　peptidoglycan, 171
　　spore coat, 171
　surface tension and hydrostatic forces, 167–168, Fig. 9-2
　phage self-assembly, 18–21
　peptidoglycan, role, 163–166, Fig. 4-5
　polysaccharide, role, 172, Fig. 9-3
　ribosome assembly, 168
　shape determination and maintenance, 163–166
Motility, myxobacterial:
　adventurous ("A"), 206
　genetic approaches, 206
　social ("S"), 206
Multicellular development, 5
Myxobacteria:
　adenylate energy charge, in different developmental cell types, 143–144
　adventurous ("A") motility, role, 128
　aggregation, 110, 132, 136–138, 140–141,

Myxobacteria *(continued)*
　144, 147, Figs. 7-3, 7-4, 7-18
　autolysis, role of development in, 142
　bacteriophage, 130, Table 7
　　Mx1, 130, Table 7-1
　　Mx4, Table 7-1
　　Mx8, 130, Table 7-1
　　Mx9, Table 7-1
　bacteriophage Mx1, development and, 130
　bacteriophage Mx8, lysogeny, 133
　bacteriophage, *Stigmatella aurantiaca*, 130
　cell density signal, adenosine, 137
　cell interactions, 200–203
　　cell density effects, 200–202
　　extracellular complementation, 200–201
　　myxobacterial hemagglutinin, 200, 202
　　social ("S") motility, 200–201
　　Stigmatella pheromone, 200, 202–203
　cell structure, 119–123
　　myxospores, 120–123, Fig. 7-14
　　vegetative cells, 119, Figs. 7-11, 7-12
　chemotaxis, 129–130; theoretical difficulties, 129
　Chondromyces, 111, Figs. 7-6, 7-9
　coliphage P1:
　　infection by, 130
　　transduction by, 130
　　transposon Tn5, 130
　colony morphology, phase variation, 133
　cultivation, 118–119; nutritional requirements, 118
　derivation of the word, 123
　development, 136–146
　　ADP, effect of, 139
　　aggregation, 110
　　autolysis, 110
　　cAMP, effect, 139
　　colonial morphogenesis, 137
　　conditions for, 110
　　extracellular complementation, 146
　　guanosine tetra- and pentaphosphate, role, 138
　　high cell-density requirement, 110, 137
　　marker events, 138–141
　　morphological events, sequence, 139, Fig. 7-18
　　nutrient depletion, 110
　　nutritional signals, 138
　　protein synthesis, during, 139
　　solid surface, 110
　　spore formation, 142–144, Fig. 7-19
　directed movement, polystyrene latex beads toward, 130

Myxobacteria *(continued)*
  DNA:
    comparison to *B. subtilis* and *E. coli*, 131
    %G + C, 132
    size, 131
    small, single-stranded, repeated copies, 132
  ecological questions, 116
  ecology, 112–116
    distribution, 112–113
    high cell density, as strategy, 115–117
    life cycle, function, 114–116
    role of gliding motility, 114
    role of myxobacteria in nature, 113
    scavengers of insoluble, macromolecular debris, role as, 115
  environment, role in development, 116, 128, 137–138, 141, 145, 147, 154–155
  fruiting bodies, 137, Figs. 7-5, 7-7, 7-16
  fruiting body formation:
    conditions, 137
    timing, 138, Fig. 7-18
  fruiting body morphogenesis, polysaccharide self-assembly, role, 126
  genetics, 131–136
    DNA, 131–132
    mutants, 132–133
    P1 transduction, 134
    partial diploids, 136
    phase variation, 133
    plasmids, 135–136
    recombinant DNA, 136
    tandem duplications, 136
    transduction, 133–134
    transposon, TN5, 134–135, Fig. 7-17
  gliding bacteria, 127
  gliding motility:
    adventurous ("A") motility, 128, 129
    biophysical theories, 127–128
    function, 128
    mechanical theories, 127
    mechanism, 127–128
    social ("S") motility, 128–129
    surfactant excretion, as mechanism, 128
  high cell density, growth rate, effect on, 141
  life cycle, 110–112, Figs. 7-3, 7-4; function, 141
  motility, 127–129
  mutagenesis, 133
  mutants:
    developmental, 133
    myxospore formation, 132
    phenotypic complementation, 132

Myxobacteria *(continued)*
  myxobacterial hemagglutinin:
    role in development, 140
    synthesis during development, 140
  Myxococcus fulvus, 130, 137, Figs. 7-5, 7-15, Table 7-1
  Myxococcus stipitatus, Fig. 7-10
  Myxococcus xanthus, 105, 110–111, 118–120, 124, 129–131, 133–146, Figs. 7-3, 7-8, 7-11, 7-12, 7-13, 7-14, Table 7-1
  myxospore development:
    fruiting body formation, relation to, 144
    germination, 144–145, Fig. 7-20
  myxospore formation:
    glycerol and fruiting body spores, relation between, 142–143
    induction by chemicals, 111–112
  myxospore germination:
    high cell-density requirement, 145
    macromolecular synthesis during, 145
    orthophosphate as cell density signal, 145
  myxospores:
    comparative properties, 110–111
    glycerol induction, 112, 121, 143–145
    protein S, 122–123, 143
    spore coat, 122
    structural organization, 120
  phase variation:
    colony morphology, 133
    developmental regulation, 133
    pigmentation, 133
  pigments:
    myxobacton, 133
    phase variation, 133
  polysaccharide, 123–127
    amounts in cell, 123–124
    desiccation, protection from, 127
    diffusion of extracellular molecules, inhibition, 125
    fruiting body morphogenesis, role, 125–126
    functions, 124–127
    gliding, role, 124
    matrix for excreted hydrolytic enzymes, 124–125
    nature of, 124
    self-assembly, 126
  protein S:
    calmodulin, relation, 122
    synthesis during development, 140
    two genes for, 122–123
    recombinant DNA approaches, lacZ-Tn5 fusions, 136

Myxobacteria *(continued)*
  RNA synthesis, during development, 140
  signals, function, 200–203
  social ("S") motility:
    pili, relation to, 128–129
    role of, 128
  stable mRNA, 136, 141
  Stigmatella aurantiaca, 118, 120, 125, 131,
      137, 144, 146–148, Figs. 7-4, 7-7, 7-9, 7-16
    intercellular signals, 147
    fruiting body, 146–147, Figs. 7-7, 7-8
    myxospores, 144
    pheromone and development, 147–148
    visible light, required for development, 147
  taxonomy, 117–118
    common features, 117
    %G + C, 117
    suborders, division of, 118
  Tn5, portable, selectable marker, 134,
      Fig. 7-17
  transduction:
    by Mx4, 133–134
    by Mx8, 133
    by Mx9, 133
  vegetative cells:
    dimensions, 119
    peptidoglycan, 119
    pili, 119, Fig. 7-13
    surface layers, 119
Myxobacterial hemagglutinin, 200, 202

Newton, A., 177
Nitrogenase, 72, 77–78, 170
Noncyclic development, 5
Nucleotide polyphosphates, role in
    development:
  endospores, 154
  fruiting body formation in *Myxococcus
      xanthus*, 135, 155

Oogoniol:
  mating in *Achlya*, 197
  structure, 197
Oscillatoria, 68

Patchy peptidoglycan, myxobacteria, 119, 171
Peptidoglycan:
  Caulobacter, 57–58, Fig. 4-5
  endospores, 28, Fig. 3-5

Peptidoglycan *(continued)*
  morphogenesis, role, 163–166, 168–169, 171
  myxospore morphogenesis, role, 171
  myxospores, 171
Phage lambda:
  antiterminator, 11
  Cro, 11, 14
  lysogeny:
    autogenous circuit, 14
    cI transcription, 14
    control, 16, Fig. 2-3
    developmental process, 13–15
    mechanism of regulation, 14–15
    Q, 15
  N protein, 11, 14, 15
  promoter sites, 11
  repressor proteins, 11
Phage mu, 15–18; lysogeny, G segment, 15
Phage P1:
  lysogenic state, 13
  Myxococcus xanthus, infection by, 134
  Myxococcus xanthus, transduction of Tn5,
      134
Phage SP01, 12
Phage T4:
  morphogenesis, DNA in head, 20
  self-assembly, 19–21
Phage T7, 9
  genetic map, 11, Fig. 2-2
  mechanism of transcriptional control, 9
Phase variation, flagellar, in *Salmonella*, 15–18,
    Fig. 2-4
Pheromones:
  mating signals in yeast, 142–146; chemical
      structure, 193
  Stigmatella, 200, 202–203
  Streptococcal, 199
  water-mold mating, 196–197
    antheridiol, 196
    oogoniol, 197
Piggot, P., 40, 183
Pili:
  Caulobacter, 50, 52–53, 56, 61–62, 66,
      Figs. 4-7, 4-8
  Myxococcus, 5
Poindexter, J., 50
Polysaccharide self assembly, fruiting body
    morphogenesis, role, 172, Fig. 9-3
Population differentiation, definition, 6
Previc, E. P., 165
Proheterocyst:
  appearance, 79

Proheterocyst *(continued)*
  spacing, 80–81
Protein S, myxospores, in, 172
Pseudoplasmodium, *Dictyostelium discoideum*, 188

Raper, K., 216
Recombinant DNA, 212–214
  key elements, 212–214
  uses in developmental studies, 213–214
Rees, D., 172
Regulation of development, 175–185
  spatial events, 177–178
  timing of events, 176–177
  transcriptional control, 178–184
  transposable elements, 184–185
Reproductive cells, formation, 157
Resting cells, formation, nutritional conditions, 43–45, 110, 137–138, 153–155
Ribosomes:
  role in development:
    endospore formation, 47–48, 181–182
    Streptomyces, 101
  self-assembly, 168
RNA polymerase, specificity modification by sigma factors, 46–47
RNA polymerase modifications:
  Bacillus sporulation, regulation, 42, 46–47, 179–180
  negative:
    ADP ribosylation, 12
    phosphorylation of β' subunit of polymerase, 12
    T4 and T7, 12
  positive, phage SPO1, 12

Self-assembly:
  bacteriophage, 6, 19
  phage base plate of T4, 19–20, Fig. 2-6
  phage head of T4, 19–20, Fig. 2-6
  T4 morphogenesis, experimental strategies, 19, 20
  tail fibers of T4, 19–20, Fig. 2-6
  tail of T4, 19–20, Fig. 2-6
Shapiro, L., 177
Shockman, G., 165, 218
Sigma factors:
  Bacillus sporulation, 42, 46–47, 179–180, Fig. 10-2
  Caulobacter, 181
  myxobacteria, 181

Sigma factors *(continued)*
  RNA polymerase modifications, role in endospore formation, 46–47
  Streptomyces, 181
Signals:
  aggregation in *Dictyostelium discoideum*, 188–192
  autoinducer of bioluminescence, 199
  bioluminescence, autoinducer, 199
  cAMP, 188
  chemotaxis, 188
  forms of, 186–187
  mating, in yeast, 192–196
  myxobacteria, 200–203
  pheromones, streptococcal, 199
  prokaryotes, 197–203
  steroid hormones, in water molds, 196–197
  streptococcal pheromones, 189
Simon, M., 15, 184
Social ("S") motility, myxobacteria, 200–201
Sonneborn, T., 4
Spatial aspects of development:
  membrane-nucleoid association in *Caulobacter*, 178; ribosome-mRNA complex, 178
  organizational centers in *Caulobacter*, 177
  polarity of *Caulobacter* cell membrane, 177
  polar structures of *Caulobacter*, 177
  temporal development in *Caulobacter*, relation to, 178
  vectorial metabolism, 177
Spore coat, 28–29
Spore cortex, 28; peptidoglycan, structure, 29, Fig. 3-5
Stable mRNA:
  endospore formation, 182–183
  flagellin A synthesis in *Caulobacter*, 61
  Myxococcus xanthus, 182
  S-protein synthesis in *Myxococcus xanthus*, 141, 182
Stanier, R. Y., 68, 205
Stent, G., 219
Steroid hormones, mating signals in *Achlya*, 196–197
Streptococcus faecalis:
  mating signal, 199
  pheromones, 199
Streptomyces:
  antibiotics, 93–96
    ecological role, 94
    functions, 93–96
    regulatory/developmental role, 95

Streptomyces (continued)
 development, 85-104
  "A factor," 103-104
  "C factor," 104
 DNA, 102-103
  %G + C, 102
  inverted repeats, 103
  size of genome, 102-103
 extracellular factors, 103-104
 fine structure, Figs. 6-6, 6-7, 6-9, 6-10, 6-11
 genetic map, 97-98, Fig. 6-12
 genetics, 96-100
  bald mutants, 97, 193, Figs. 6-13, 6-14
  epistasis experiments, 97
  plasmid transfer, 97
  transformation, 97
  "whi" mutants, 97, Figs. 6-13, 6-14
 regulation of development, 100-102
 ribosomes, 101
 rifampicin-resistant mutants, development, 101-102
 spore formation, macromolecular synthesis during, 100-101
 spores, 87-93
  fine structure, 87, 89, Fig. 6-7
  germination, 91-93
Sussman, M., 216

T-even bacteriophage, growth cycle, 10, Fig. 2-1
Thaxter, R., 105, 117
Thompson, D'Arcy, 163
Timing of developmental events:
 Caulobacter, 177
 cell division in Myxococcus xanthus, 176
 DNA replication as cellular clock in Caulobacter, 177
 genetic programming model, 176-177
 metabolic model, 176
Tn5, 212-214
Transcriptional control:
 bacteriophage development, 9-13
 glutamine synthetase gene in Anabaena:
  promoter sequence, 181
  regulation, 180
 RNA polymerase of:
  Caulobacter, 181
  Myxococcus xanthus, 181
  Streptomyces, 181
 sigma factors of:
  Caulobacter, 181
  Myxococcus xanthus, 181

Transcriptional control (continued)
  Streptomyces, 181
 Transcriptional control of development:
  Anabaena, 180; multiple promoters, 180
  Bacillus sporulation, 46-47, 179-180
   overlapping promoters, 180
   nutritional control of spo gene expression, 179
   sigma factors, 42, 46-47, 179-180, Fig. 10-2
   spo genes, 179-180
   RNA polymerase specificity, 42, 46-47, 179-180
  strategies for bacteriophage development, 9-13
 Translational control of development:
  erythromycin resistance in B. subtilis, 181-182
  ribosomes of Bacillus subtilis, role, 182
  sporulation and germination in Bacillus subtilis, 182
  stable mRNA in Myxococcus xanthus, 182
 Transposable elements:
  control of development, role, 184
  phase variation in Salmonella, 184
 Transposons, 205-206, 212-214
  developmental genes, genetics of, 212-214
  selectable markers, 212-213
  Tn5, 212-214

Unicellular development, 5

Water-mold mating, signals, 196-197
White, D., 147, 171
Whittenbury, R., 4
Wolf-pack effect, in myxobacteria, 141
Wolk, P., 83, 170
Woods, N.A., 105
Wright, B., 176

Yeast life cycle, Fig. 11-3
Yeast mating:
 cassette model, 195-196
 HO locus, 195
 MATa, 195-196
 MATα, 195-196
 mating types, 193-196
 mating type switches, 193-196; mechanism, 194-196

Zusman, D., 213